T0138533

Elton's Ecologists

Peter Crowcroft

Foreword by Thomas Park

Elton's Ecologists

A History of the Bureau of Animal Population

The University of
Chicago Press
Chicago
& London

PETER CROWCROFT, an alumnus of the Bureau of Animal Population, is lecturer
in zoology at the University of Texas at Austin. He has served as director of
several zoos in North America and his native Australia, including Chicago's
Brookfield Zoo, Metro Toronto Zoo, and Taronga Zoo, Sydney. His books include *The Zoo, Mice All Over, The Life of the Mole,* and *The Life of the Shrew.*

The University of Chicago Press, Chicago 60637
The University of Chicago Press, Ltd., London
© 1991 by The University of Chicago
All rights reserved. Published 1991
Printed in the United States of America
00 99 98 97 96 95 94 93 92 91   5 4 3 2 1

Library of Congress Cataloging-in-Publication Data

Crowcroft, Peter.
   Elton's ecologists : a history of the Bureau of Animal Population
/ Peter Crowcroft.
        p.   cm.
   Includes bibliographical references and index.
   ISBN 0-226-12146-1 (alk. paper). —— ISBN 0-226-12148-8 (pbk. :
alk. paper)
   1. University of Oxford. Bureau of Animal Population—History.
2. Animal ecology—England—Oxford.   3. Elton, Charles S. (Charles
Sutherland), 1900–   .   4. Ecologists—England—Biography.
I. Title.
QL69.G72U553   1991
591.5'071'142574—dc20   90-39469
                                                                   CIP

⊗ The paper used in this publication meets the minimum requirements of the American National Standard for Information Sciences—Permanence of Paper for Printed Library Materials, ANSI
Z39.48-1984.

Contents

Foreword

Occasionally, very occasionally, a unit emerges within a larger academic structure and, through time, endows a field of inquiry with new orientation, new meaning, and an expanded understanding of its own conceptual significance. In short, intellectual quality is progressively created. Typically, this quality evolves, not from prospective planning, but rather from the dedication and talents of a few participants. Usually, also, there is one key figure who, irrespective of methodology, is primarily responsible for the success of the enterprise. This book by Peter Crowcroft relates such a story. The place is the University of Oxford, the unit is the Bureau of Animal Population, the field is ecology, and the protagonist is Charles Elton, F.R.S.

In the pages that follow, Peter Crowcroft, himself both a graduate and a friendly observer of the Bureau of Animal Population, tells the story of this smallish university department. The story, engaging in its own right, chronicles the birth, maturity, and death of an enterprise that appreciably shaped the history of ecology during the mid-decades of the twentieth century.

Much of contemporary science is programmed, compartmentalized, firmly administered, and often copiously financed. There are patent and compelling reasons for this approach; the rewards, both intellectual and technical, accumulate with impressive regularity. However, it is pleasant to remember, while memories still linger, what can be called "small science"—science that has been practiced in a sometimes faltering but often adventuresome way. *Elton's Ecologists* documents just such a case history.

Thomas Park

Preface

Elton's Ecologists is the collective label I have applied to the people who worked for months or years in the Bureau of Animal Population, Oxford University, England. The BAP was a small research institute founded in 1932 by Charles Elton and directed by him until he retired, and it ceased to exist, in 1967. If you have not heard of him you are unlikely to be reading this book.

Writing this history of the BAP has been my self-imposed intellectual burden for about fifteen years. All too often, I have felt the task hung about my neck like the Ancient Mariner's albatross. But again, just as often, I have been glad to have something worth doing to turn to after a day as a zoo director, a day spent dealing with public relations, union grievances, or hot-dog sales. More important, the book gave me an excuse for visiting Charles Elton in retirement and getting to know him better. As he moved into his 80s, he slowed down just enough to enjoy talking with me without getting restive.

I have never been able to do my job well and care properly for my family at the same time. That is common enough, but few men in that predicament are rash enough to take on the writing of a book, which soon makes more demands than a mistress. Two quite different reasons prevented me from giving up: I would have been ashamed to admit to Elton ("The Boss") that I had found the task too difficult, and I had an excessively sentimental attachment to my own association with the Bureau. It is probable that working with Elton and his colleagues changed my attitudes, even my life, more than those of other students. That provides some excuse for my self-regarding intrusions into the general history.

My albatross was shot on a wet afternoon in Oxford in 1972. At that time I was managing the Brookfield Zoo in Chicago and my employers were very generous in regard to overseas travel. I was on one of my regular visits to Britain to see my sons, who were at Canford School in Dorset. As I turned my rented car on to the roundabout in North Oxford where the Woodstock Road crosses the A-40 I glimpsed a street sign, "Five Mile Drive," and on impulse I decided to call on Denys Kempson, the retired inventive genius of the BAP. "DK" would be pleased to see me, but embarrassed at my turning up without telephoning; DK had always been friendly and helpful, but he had tended to keep his professional and private lives apart. I needed a reason to call on him, and as I drew into the kerb at No. 11, I was reflecting on what a pity it was that so few people knew how much DK had contributed to the Bureau's productivity. By the time DK had opened his front door, taken his pipe from his mouth, and exclaimed, "PC! What brings you to this part of the world?," I had an answer ready.

"DK," I said, "I need your help with a little book I'm planning to write about the Bureau." I explained that I proposed to select some photographs from the Bureau's archives. If I got the Boss's permission, would he seek out the negatives and make me some good prints? Naturally, I would expect to pay for any expenses involved. DK looked very grave as he said that he liked the idea but would have to get the blessing of "The Chief." I assured him that without Elton's approval and cooperation the book could not be written; I proposed to put together an outline and submit it to the Boss for his reaction.

On the way back to London from Canford I called to see Charles Elton and dropped the idea on him. As I expected, he was very wary and reserved. He loathed personal publicity and there was also a natural reluctance to discuss the demise of the BAP, which ceased to function on the day that he retired. After thirty-five years of successful operation and international regard, the BAP had been done away with in spite of his best efforts to preserve it. He wasn't very happy with the idea of a book but he agreed to look over a synopsis and rough outline of the treatment. There was no reason to hurry and I felt he might like the idea better when the demise of the Bureau had receded further into the past. A few years' delay happened because I accepted a proposal to write a book about zoological parks. A cash advance was offered to me just when I needed funds to travel to a conference on the captive breeding of endangered species. I had foolishly left the Brookfield Zoo to manage Taronga Zoo, Sydney, and my employers there were under the thumb of a state premier who objected to the overseas travel of other people.

The delay turned out to be for the best. Two factors produced a more positive attitude in the Boss. One was an attempt by a graduate student to convert his thesis into a book which would be focused on Charles Elton and the BAP. Elton did not like its tone, especially certain assumptions about his personal thoughts and motives. The second reason was that he very much liked the book I wrote about zoos (25). He decided that if anyone was going to write a book about the BAP, then I was the person who should do it. And to make sure I wrote an accurate account he proceeded to give me much more help than I had expected. Other former Bureau people also became more helpful because of my book *The Zoo*. My former supervisor, H. N. (Mick) Southern paid me the backhanded compliment of saying that if I could make a dull subject like zoo management sound interesting I could probably do the same for the Bureau's history. Alas, I soon moved my tent again because I came to lose all respect for the zoo board in Sydney, and with the preoccupation of managing a giant of a zoo in Toronto I had little time or energy for writing.

After nearly a decade the tables were turned and the Boss was urging me to get on with the job. This was not, I learned, because he thought he might die before it was finished, but because he feared that I might. He knew a lot about actuarial tables and at the age of eighty-one he was on a pretty safe statistical plateau. He pointed out with a smile that I was, at fifty-nine, at a much more dangerous age. When I protested that there were big holes in my knowledge, and that they inhibited my writing, he reassured me saying, "Every book has holes in it." He advised me not to look over my shoulder too much or to feel obliged to submit my text to others for their opinions. I should consult others only when the facts were in question. He quoted C. P. Scott's aphorism: "Comment is free, but facts are sacred." I should also avoid writing anything that might embarrass living persons. (Those who conceivably could be embarrassed were getting fewer every year.) When I said I was sorry to be taking so long he replied with a twinkle that he had written one of his books in three weeks, another in three months, and had taken sixteen years apiece for two others.

During that conversation he showed me a little textbook on animal ecology that had just been published by a young man in the Oxford zoology department. Elton thought it was a fine job. Most books of this kind were bad, in his view, because they tended to categorize everything and to imbue untested ideas with spurious authority. He commented that animal ecology ought really to be taught as a divided subject, in three parts. First, there was the deterministic aspect; for example, "All cockroaches are brown." Second, there was the stochastic; "How many cockroaches are there in a place, and how are their numbers changing?" And third, generally overlooked, there was the historical aspect; "When did cockroaches first come to Britain?" because there were far-reaching consequences from the timing of ecological events. Then he pointed out with a wry smile that the writer of this fine book on his subject had managed to avoid any reference to C. S. Elton in the bibliography. During the tenure of Professor J. W. S. Pringle as Linacre Professor it was prudent to treat Elton as a nonperson, and to avoid any suggestion that the Oxford Animal Ecology Research Group (AERG), which Pringle had to establish in order to get his way, owed anything to its illustrious predecessor, the Bureau of Animal Population.

The book that Charles Elton wrote in three months when he was twenty-six years old was the classic *Animal Ecology* (42). The book was written, he recalled, under pressure and with a feeling of clarity. In his preface to the tenth impression he remarked with regret that it was difficult to recapture that feeling now that the complexities of the subject were more apparent. The changes in emphasis mainly concerned the complexities of habitats and of the communities living in them. A great deal of work had gone into population analysis along three lines: field studies, laboratory experiments, and mathematical models. Although too polite to use harsh terms to describe the work of others, he went so far as to say: "unfortunately, Nature cannot be understood by pretending that it is simple."

In the first edition of *Animal Ecology* Elton was intrigued by the possibility of finding a causal relationship between fluctuations in the numbers of game species and the sunspot cycle. Unfortunately, that elegantly simple explanation became unacceptable under more detailed scrutiny. The probability of an overriding effect of climate remained, but the climatic cycle was not in phase with the sunspot cycle. Many other novel ideas that became part of the conventional wisdom of animal ecology were compressed into that slender volume, or if they had been previously expressed, were modified and given new meaning. British animal ecology, except for that of oceans and lakes, in which there had already been substantial progress, began to grow in importance after the publication of *Animal Ecology* mainly because Elton was teaching the subject at Oxford. In the 1930s there were few other places, apart from Imperial College, the University of Chicago, and some Soviet research institutes, where the subject was taken seriously.

Plant ecology became a mature subject much earlier, for plants are much easier to observe, to count, and to understand than animals. Plants have to stay put and suffer being counted and can have bits cut off them for identification. Perhaps more important for ecologists is the fact that plant numbers change more slowly and for reasons that are less difficult to discern. One species may crowd out another in the course of a growing season, but an individual of one species does not suddenly arrive and gobble up a lot of others.

As well as studying a variety of animals in the field from an early age, Charles Elton had read wisely rather than widely. There were seminal ideas to be found in botanical works and in some writings about human populations. He has said that he "read a little and thought a lot." Certainly in *Animal Ecology* he made novel generalizations and placed old ideas in a new perspective. Sixty years after publication the book is not only being reprinted, it is being translated into more foreign languages. The concept of the "food chain" had been developed in fishery science and had, indeed, been recorded in art and literature before there was any biological science at all. Elton was the first to stress the significance of the jumps in the relative sizes of the species found in successive links of a food chain, and in the relative *numbers* of individuals at each stage. He introduced the term "pyramid of numbers" to sum up this principle. Each pyramid has many herbivores forming its broad base-layer, supporting one or more layers of exploiters or predators. At the apex of its pyramid there are relatively few individuals of an invulnerable species such as the lion, the crocodile or the eagle (invulnerable except to man or parasites).

The term "niche" was already in use in plant ecology, but it had a topographical meaning: a niche was the kind of physical place in which certain plants were to be found. Elton transformed "niche" into a functional concept. After the publication of *Animal Ecology* the niche of an animal came to mean its way of life; its occupation in the biological community. Animals of very different kinds, indeed, animals classified in different phyla of the animal kingdom, could occupy the same niche in different communities.

Just as an engineer can move from one construction (or demolition) job-site to another, so can a species, finding a vacancy, retain its occupation but benefit from a change of fellow workers. When a species is transported to another country, usually but not always through the actions of man, it can find such unbounded opportunities in its new world, because its niche has been unoccupied, that it can become highly successful. If it finds a niche that has been unoccupied through zoogeographical barriers or accidents of history, and is open for any candidate with the right attributes for doing the job, it can undergo a career change, and become so successful it comes to man's attention as a pest. Tracking down the records and history of plant and animal pests became a special interest of Elton's, resulting in his book *The Ecology of Invasions by Animals and Plants* (50). In this book, as in other works, Elton made generalizations, which although clearly stated for the first time, were so evidently true that readers were apt to comment, "Well, yes, but that's obvious isn't it?"

I have assumed, thus far, that my reader knows what "Ecology" means. But the word has been so adapted and misused by special-interest groups and politicians, and even by commercial firms (Phillips once advertised an "Ecologizer" for use in bathrooms) that its meaning has become fuzzy. Frank Fraser Darling made the gloomy comment, "By the time people know what 'Ecology' means, it will be too late." That was during an international conference on the ecological consequences of "development" in various countries. The function of the meetings was to review and comment on the unforeseen and usually disastrous consequences of making major changes to natural systems, such as building the High Aswan dam on the Nile. An Egyptian scientist was there to report on the loss of coastal grain-growing land due to the absence of silt previously discharged from the mouth of the great river. He had been fired because his report to his government lacked enthusiasm. The great economist and friendly critic of the United States, Gunnar Myrdal, was there, and a host of heavyweights in various disciplines. Raymond Dasmann, then working for the International Union for the Conservation of Nature had invited me to relate an example of ecological folly in Australia (24) and partly, perhaps, because we had shared hard times on the Zambezi at Kariba, another African folly.

As the conference proceeded, the rhetoric increased and the term "ecology" was being misused in so many ways that I felt I had to speak up. At one session Barry Commoner was in full spate about "the uniquely holistic approach of the ecologist" to the planning of great projects. When he eventually paused for breath I quickly rose and commented to the chairman of the session that the word "ecology" was being dragged back and forth across the path of the conference like a smelly red herring. The word did have a specific meaning and we would be well advised to stick to it if we wanted politicians to take us seriously. It was foolish to say that we would bring to the planning table a uniquely holistic approach; others such as architects, economists, and geographers would make the same claim, and with equal validity. What we had to offer that was unique, was an informed opinion about the consequences of changing habitats. Ecology was the study of organisms in relation to their environment, so when major environmental changes

were going to be made, ecologists were the people best able to predict
the likely consequent changes in animal populations.

After an awkward pause, Barry proceeded as before. But at the coffee-
break a distinguished-looking gentleman sought me out. He took a pen
and a used envelope from his pocket and asked, "Would you mind telling me again what ecology is?" He carefully recorded the definition,
thanked me, and strode off. He had not introduced himself, so I pointed
him out to Ray Dasmann and asked who he was. When Ray told me his
name and occupation I asked, "What on earth is he doing at a conference like this?" "Well for one thing," said Ray drily, "he's mainly responsible for us having the conference. You see, he's President Nixon's chief
advisor on the environment."

Much of the woolly thinking so prevalent in discussions about ecology is due to failure to distinguish between ideas and techniques.
Whether or not someone is an ecologist is determined, not by what he
does, but why he does it. A piece of research may involve using techniques developed in other disciplines, and the use of them need not imply any relevance to the ideas which caused them to be developed. For
example, using electronic gadgets to measure temperatures in soil or litter does not make the investigator a physicist. Similarly, studying living
things in the field does not make the observer an ecologist; he or she
may be an ethologist, a forester, or a geographer. Elton tends to define
ecology in terms of the study of *populations* in relation to the environment, as he has been interested in dynamic processes. Other eminent
ecologists, from pioneers like Royal Chapman to some present-day
teachers, would not define their subject in the same way. (Although
populations have now come to the fore through the amazing development of techniques for genetical analysis.) But all would agree that the
basic drive of the ecologist is to ask *why,* rather than *what.* Ecologists are
the modern counterparts of the nineteenth-century naturalists, but they
(mostly) have a wider perspective than (most) of the old-fashioned students of natural history. Nowadays one does not ask what animals are
present in a particular locality, and to which other animals they are related. Now we want to know why a species is found where it occurs and
not in other places; why there are more (or less) individuals than last
year. Being interested in numbers does not necessarily involve the use of
sophisticated equipment, but it does involve much more sustained effort.

Understanding changes in the numbers of animals is necessary for
managing populations of those species we decide to save from extinc-

tion, and for saving our own species too. But most ecologists are not investigating some aspect of the natural world because they want to save the diversity of life; they are studying something that has stimulated their interest and curiosity. Many are seeking answers to specific questions asked by their supervisors. Very few have the breadth of experience or the intellectual capacity to put forward new ideas of a general nature. Elton has become recognized as one of the fathers of modern ecology mainly because he dissipated the foggy but seductive notion of a "balance of nature" and replaced it with the idea that nature is constantly in a state of change, with the numbers of a species in a place changing from year to year as well as from season to season. Others have been involved in this revolution of thought, but Elton's book *Animal Ecology* stands as a marker between the old natural history and the new.

Charles Elton has spent most of his life gathering and weighing facts that might throw light on the processes that keep species from increasing beyond the levels they tend to fluctuate about, or from disappearing altogether. He has sought to formulate principles but has not regarded them as laws such as those that consistently define how chemicals will react when brought together or how an electric current generates a magnetic field. He has stressed that ecological principles have lots of exceptions to them, and much depends upon what is happening locally; on what is going on in a particular place at a particular time. For a person who has been reluctant to press his ideas on anyone, except through his writings, and as that rare kind of leader who shuns personal promotion and publicity, Elton's influence on animal ecology has been remarkable.

On hearing of my proposal to write a history of the BAP, Robert Rudd, a veteran teacher of zoology in a great university wrote to me: "I don't suppose Elton has a full appreciation of how much impact his name has. . . . I never met him but I've heard the name since I was very young. Many of us are admirers, and certainly many more latter-day students are Eltonian thinkers sub-liminally." Another wrote to me: "I was always impressed, and remain so to this day, with the productivity and influence of the Bureau in shaping modern Ecology. It was, after all, a small unit which existed on a minuscule budget, yet it produced outstanding research and a group of students who have assumed positions of leadership around the world." Such remarks led me to plan to publish a map showing where the BAP's graduates came from, and where they worked at the time of publication of this book. But writing this account took so long that most of them had retired or died. As I wanted to publish a celebration of the Bureau's life and times rather than an epitaph, I gave up the idea. As I write this introduction for a book that is almost completed, Charles Elton is eighty-nine years old and likely to outlive all of his students as well as his contemporaries.

But as the self-appointed president of the Elton fan club I find I head a small constituency. The end of the Bureau is now many student-generations in the past. When I mentioned the Boss by name to a group of graduates in wildlife management a few years ago they thought I was talking about Elton John. And when I reacted testily, "No, no, not Elton the singer, Elton the ecologist," one of them commented, "Oh, I didn't know Elton was into ecology." "But haven't you heard about niches and pyramids of numbers in your course-work?" I persisted. Then another of them said, with clear relief that he could satisfy this demanding geriatric case, "Oh. You mean all of that stuff in Odum's book!"

When the Boss reads this book he will like that story, just as he liked Eugene Odum and his highly successful textbook, especially the second edition (115). The first edition he reviewed in the *Journal of Animal Ecology* in 1954, and expressed the reservations he had about all authoritative books: "It is the absolute necessity for students to read them as critically as they would a special paper or monograph. Animal ecology is still in such an embryonic stage of thought (though it has all too many facts to swim about in) that it can hardly be said yet to have completed its neural fold. The facts are rather chaotic; many of them are not facts at all; its theories are poised uneasily between arm-chair pipe-dreams (valuable as models for thought) and ready-to-wear mathematical models that fit badly and are already bursting at the seams. And we have still not solved satisfactorily our chief dilemma—how to study the full complexity of interlocking animal communities without sacrificing depth to breadth of research." He was a bit put out by Odum's initial failure to recognize that Elton's use of the term "niche" in 1927 was fundamentally different from previous usage: "The author does not make it clear that Grinnell's concept of niche was different from mine, in emphasizing the distributional and spatial position of a species, whereas I used the notion of a functional place in a dynamic community—mainly but not exclusively in regard to food habits." He especially liked Odum's chapters on population structure, and strongly recommended the book. Nowadays, as a teacher, I find that most students from other universities have not heard of Elton *or* Odum! But then, many of them have not heard of Julian Huxley or J. B. S. Haldane either.

Over sixty years have passed since Elton's concept of ecological niches started to illuminate thought about the composition of animal communities. His was a simple, broad concept, and useful because of that. Attempts to develop quantitative and multidimensional niche theories have tended to obscure its clarity. The initiative of ecologists skilled in algebra has provided material for many theses and publications, but has advanced understanding of the natural world about as much as theological debates about the size and morphology of angels. Many research projects have a social function similar to that of theological studies in past centuries; they provide work and career advancement for people not interested in the pedestrian work of a parish. Charles Elton has combined the inspiration of a theologian-ecologist with the dedicated hard work of a parish priest; he is a competent taxonomist in a number of difficult groups and an indefatigable collector of specimens and associated field data. In retirement, he has been more concerned about the preservation of his Wytham Biological Survey material than about the survival of anything he has had to say about ecology. This was brought home to me when I visited him in the Oxford zoology department in 1987.

The Department of Zoology is now housed with the Department of Experimental Psychology in a concrete monument to an architect's cleverness embedded in the corner of South Parks Road and St. Cross Road. Its exterior is no more displeasing than that of other postwar Oxford buildings. Inside, it is a maze of interlocking split levels and stairways like the scaled-up version of a rat-conditioning maze. Much of the first floor is devoted to very wide corridors for the rapid egress of lecture-hall audiences. On the second floor there is a spacious gathering place where low sofas invite the unwary to sit and crack their heads on the unfinished concrete walls of the building. Groups of chairs are arranged in the center to encourage the formation of temporary discussion groups at "tea time," and these are used in the socially stratified manner still customary in Britain; faculty sit together and so do technicians, labeled by their white coats. The librarians sit near their library door, making a dutiful appearance, after which they can return to the library and relax. Students can and do approach the teaching staff, and they can kneel quite comfortably on the carpet.

Further upstairs the scale is different. There are cramped corridors between rows of small cellular labs and offices. Each cell has its share of those intrusive ducts and mysterious protuberances not shown on the original plans. In one of these I sat knee to knee with Charles Elton. Most of his space was occupied by museum cabinets containing the thousands of specimens collected in Wytham Estate, and his desk was covered with new record cards inscribed with his neat, minute script in India ink. He told me he was not happy about his continued presence in the department, so long after his retirement. He did not want to become "the old chap who wrote *Animal Ecology* sixty years ago," haunting the department. But he was very concerned for the safety of his material and records. Without his unpredictable presence, he felt they would be squirreled away in a basement and forgotten. Then, in another decade, with another departmental head, they might be thrown out as rubbish. I could sympathize with him as I had saved material from the scrap heap while working in the British Museum, and seen it rehabilitated as precious, irreplaceable material. He said that he would value my opinion. It took only a moment to reply. Trying to sound flippant, I said that he should endeavor to keep interest alive in the Wytham Biological Survey through his personal stewardship, but that if he found it a burden, I felt he could review the decision, with a clear conscience, on his ninetieth birthday. He appeared to find that answer acceptable.

Trying to retrace my steps to South Parks Road, I passed two graduate students in lively conversation. I loitered and pretended to be examining some of Alister Hardy's lovely illustrations of marine life, which then adorned the wall of a corridor. They were discussing an obscure theoretical point in ecology. As they fenced and scored, each showing the other how clever he was, their argument lifted above the ceiling of my comprehension. My gloomy mood, brought on by the contrast between the working atmosphere of that place and the warm friendly old Bureau, was suddenly lightened. The words with which Lord Tennyson sought to rationalize the defeat of the good guys from Camelot came to mind:

> The old order changeth, yielding place to new,
> And God fulfills Himself in many ways,
> Lest one good custom should corrupt the world.

These bright young things would not be dulled by the concrete and congestion. The Bureau of Animal Population had provided a wonderful environment in which a few local and exotic varieties had been able to grow. The BAP was no more, but animal ecology and Charles Elton were both alive and well in Oxford.

I had the good luck to work in the BAP for some years and the good sense to remain a hanger-on for most of its life. But as I have spent most of my career managing scientific (and pseudo-scientific) institutions rather than in more intellectual work, what follows has to be prefaced with a quotation from Francis Ratcliffe's book, *Flying Fox and Drifting Sand* (122).

> I hope that no one will be tempted, because a certain importance attaches to the subject of my scientific wanderings, to judge this book in a class to which it does not pretend. It is essentially a collection of observations, impressions, and reminiscences, on the whole more subjective and trivial than scientific and serious.

1   The Birth of the Bureau

Elton is not an uncommon name. There was a crusader named De Acton, and in later generations the name became De Elton and then plain Elton. There are four villages in England called Elton, and Thomas Park, the renowned Chicago ecologist, once sent Charles a postcard from Elton, Wisconsin. Oxford folk often asked if Charles was related to Jack Elton, who kept a butcher's shop in North Parade, and he probably was, for there was a long line of gentleman farmers, yeomen of England, who had sired many sons. Charles's second given name, "Sutherland," had its origin in Professor Oliver Elton's felicitous habit of naming his sons after professorial colleagues. Charles S. Elton is very proud of his descent, on his mother's side from crofters on the Isle of Coll, Scotland. About 1933 he mentioned to his class that to get a perspective of nature they really ought to visit a small island and explore it thoroughly. That year eleven of them went on island expeditions, and it gave him special pleasure that one of them, Jean Taylor (who later became Mrs. Peter Medawar), went to Coll.

Charles was born on March 29, 1900, in Manchester, where his father held the chair of English literature at the university. Professor Elton moved to Liverpool University, and it was in the vicinity of Liverpool on England's east coast, that the young naturalist developed, under the guidance of his older brother, Geoffrey Yorke Elton, from the age of nine until he went to Oxford University as an undergraduate. In 1918 Charles was head boy at Liverpool College, which his Oxford contempories from the great public schools liked to style "a minor public school in the North." He spent four months in the Army Cadet School (Royal Engineers-Signals) preparing for army service, but the "war to end all wars" ended in November of his eighteenth year, before he could be consumed.

Elton's tutor at Oxford was another brilliant young naturalist, Julian Huxley, grandson of the Darwinian apostle Thomas Henry Huxley. Julian was sufficiently impressed by one of Charles's essays to suggest that they should work it up into a joint publication. But the young Elton was already his own man, and this suggestion, which most undergraduates would have embraced with suitable humility, elicited a cool, "No, thank you. I am intending to send it as a Letter to *Nature*." A certain degree of tension seems to have persisted in their long relationship, for after one of Huxley's visits to the Bureau, Elton came back into the library, eyes flashing icy blue sparks, and dropped into a chair remarking, "I don't really mind when Julian steals my ideas. But I strongly object when he takes out a notebook and writes them down in front of me." This was a moment of great disillusionment for me, as I had read Huxley's popular writings in my teens, and had placed him on a very high pedestal indeed.

Even in his undergraduate years at Oxford, Charles Elton was developing original and novel ideas about the processes of natural selection, and attempting to test them by observing animals in the field. (39). If he had not become mainly interested in population dynamics, he would certainly have achieved eminence in evolutionary theory. Some of his ideas about natural selection were the foundation of a series of three lectures on "The Future of Animal Ecology" given at University College, London, in the fall of 1929. They were published in 1930 as *Animal Ecology and Evolution* (43). The book did not attract much notice, but Alister Hardy was impressed and drew attention to Elton's contribution in some of his writings (69, 70). Like Huxley, Elton had watched the behavior of Great crested grebes and thought about its significance, but he did not follow up that subject either.

Those novel ideas are revived in his last book, *The Pattern of Animal Communities* (51). Animals do not sit about waiting for the environment mindlessly to select the fittest to survive, as plants must. They generally can and often do move from places in which they are not doing well into places in which they may do better. This amounts to *natural selection of the environment by the animals.* When the migrations of animals are taken into account, the range of environmental conditions with which the inherited variations in a species can interact must be greatly increased.

The influences which focused Elton's attention on the *numbers* of animals, and on the composition and changes in animal communities, have been explained by him in a talk he gave at the Bureau on February 14, 1962. He disliked getting up and speaking in front of a crowd of people, but he gave this account of the first thirty years of the BAP's work because of Alister Hardy's retirement from the Linacre Chair of Zoology, and imminent retirement from the Chair of Zoological Field Studies, of which the BAP was a part. I was given a copy of this unpublished talk to make sure I got my facts straight. It was also to avoid being exposed to interrogation on the subject. Elton told me he trusted my judgment in making use of it, and at the same time made my task almost impossible by asking that he should not figure too prominently in my history of the Bureau of Animal Population!

In 1920, while an undergraduate, Charles Elton was strongly influenced by Victor Shelford's book, *Animal Communities in Temperate America* (124). This contained an account of an *ecological survey* carried out in the vicinity of Chicago, Illinois. Just when this interest in surveys was growing, Elton had the opportunity to go to a place so inhospitable for living things that a survey could be fairly complete; Julian Huxley invited him to be his assistant on the first Oxford Expedition to Spitsbergen in 1921. Huxley let him do as he chose, and he chose to undertake an ecological survey of the Spitsbergen fauna. "Old Warden Spooner once remarked to Huxley in his quaint way: 'I understand that Spitsbergen is no further from the North Pole than Land's End is from John of Gaunt!' It is heavily glaciated, ice-bound for most of the year and some parts even in summer, and because of its high latitude and geographical isolation from Eurasia has a very limited fauna. When I had finished reading a paper on the very sparse beetle fauna of the archipelago, at the Entomological Society of London, I heard Professor Bateson, the great Cambridge geneticist, remark in a stage whisper: 'I don't think these Oxford men know *how* to collect beetles!' In fact only six species were known."

This was the first of three expeditions organized by George Binney, who was later to become an important officer of the Hudson's Bay Company, and a supporter of Elton's survey work in North America. The second expedition was the Merton College (Oxford) Arctic Expedition of 1923. This time the students tried to circumnavigate North East Land, "and were led to some very barren spots indeed." On the whole of the island, which is about the size of Wales, Elton found only nine species of dry-land invertebrates out of the sixty or so recorded from the entire archipelago. A most significant incident was that on the way home to Britain Elton happened to spend one of his last three English pounds in a bookshop in Tromsø. He bought Robert Collett's book on Norwegian mammals, *Norges Pattedyr* (21). In that book he learned about lemming migrations. He got himself a Norwegian dictionary, translated the records, mapped the occurrences, and found that the migrations had occurred somewhere in Norway every three or four years.

Elton was in charge of the scientific work of the third expedition, the Oxford University Arctic Expedition of 1924. Much of his time was spent in the base camp on Reindeer Peninsula, and from there he completed the general survey as well as he could (41). He nearly came to an early end when he fell through the ice up to his neck, an accident which may have stimulated his thoughts about the importance of accidental occurrences in population dynamics. (The importance of accidents received little attention in theoretical ecology until they were restyled *hazards* by Browning [7], and then classified as *malentities* by Andrewartha and Birch [1]). The Spitsbergen expeditions resulted in the publication of a number of papers in botany and geology as well as zoology. Probably the most important were those on the vegetation, by Summerhayes (and Elton) (137, 138). Although Elton's name appeared as junior author, he carried out most of the research, as Summerhayes was not a member of the second and third expeditions.

In relation to the future course of events, the most important outcome did not concern the completed scientific studies. For two weeks, Elton shared a tent with a young Australian named Howard Florey, who was medical officer on the third expedition. Florey went on to become famous for developing the production and use of penicillin, and became one of the very few Australians, apart from retired prime ministers and governor generals, to enter the House of Lords. It was Florey who suggested to Elton that he might well utilize the talents of P. H. Leslie in his search for the causes of vole cycles. No doubt Elton's impression of Florey reinforced his liking for "colonial types," a liking which began at New College, where there were many Rhodes Scholars from the United States, Canada, South Africa, and the Antipodes. As this was just after the Great War, these students tended to be more mature than usual, and they were in Elton's view "a lively and very intelligent bunch."

Two other books affected Elton's thinking at this time. One was *The Conservation of the Wild Life of Canada*, by Gordon Hewitt (72). It contained graphs of the annual fur returns of the Hudson's Bay Company, "two of the most striking being those for the Snowshoe rabbit (or Snowshoe hare) and Canada lynx. These could be seen to fluctuate in a remarkably regular manner and to have a periodicity of about ten years." In the lemming migrations and the fluctuations of the Lynx and its staple prey, Elton saw two aspects of what might prove to be a general phenomenon. The other book was *The Population Problem*, by Oxford professor Carr-Saunders (9). Although this was about human populations, "it was full of exciting ideas of a general nature." Elton devoured this book two weeks before his Schools (the final honors exams at Oxford). He says he was not very good at doing formal exams and must have got first-class honors because of the thesis he submitted on the Spitsbergen work.

With a first in zoology he could have followed the usual academic career track by taking a teaching and research post in a lesser university. Instead, he decided to remain in the Oxford department of zoology and comparative anatomy. He would work on the investigation and documentation of fluctuations in the numbers of animals, and watch for clues to underlying principles. In order to have a place to work and some income for essential living expenses, he accepted the post of departmental demonstrator in zoology. It was only a temporary, part-time position, but he was successful in getting a number of modest grants to support his research. He was soon promoted by Goodrich to the permanent post of university demonstrator, and he remained in that job as "a thread of teaching for a small retainer," from 1929 until the end of the 1939–45 war. Elton's research proceeded along two lines of inquiry. It might be better to say that he attacked the one general problem from two directions. One approach was through a system of special reports from the far-flung empire of the Hudson's Bay Company in North America. He had been appointed biological consultant to the company in 1925, and was retained in that role for five years. Knowledge of the causes of drastic fluctuations in the numbers of fur-bearers, especially if that knowledge led to predictions of changes, would be of great practical use to the industry. George Binney gave him administrative help in putting his recording system into place. From their personal observations, local trading-post officers recorded trends in the numbers of certain animals. Were they more abundant or less abundant than last year? A simple judgment to make, but when made by hundreds of observers who knew what they were writing about, the information was capable of objective analysis and it became more valuable as the years passed. The animals selected for the report were not only the fur-bearers of commercial importance, but also their prey, and some other species whose abundance or scarcity might confirm the general phenomenon. The wide geographical range of the reporting stations could be expected to reveal any synchronization which might occur between the changes in different populations. When Elton later set up his own research institute one of its main functions was to service this reporting system, analyze the data, and regularly publish the results (12, 13, 46). The company's archives enabled the investigation to be carried back in time. From the massive bulk of old paper it proved possible to trace the ten-year cycle in the abundance of the Canada lynx back to 1736 (57).

The second line of attack consisted of investigating fluctuations in the numbers of British voles (*Microtus* and *Clethrionomys* and field mice (*Apodemus*). Elton might have preferred to study mammal populations in Labrador or Lapland, but it was not practicable. The small mammals which could be studied using Oxford as a base had also been reported to fluctuate in numbers (48). The British work got started after a conversation between Charles Elton and John Baker during a morphology demonstration. Elton said something like: "I say, let's do something with mice." And Baker responded: "Jolly good idea. I can contribute my car." They recruited another young colleague, E. B. Ford, forming the team which became known affectionately as the Mouse Gesellschaft. Baker undertook to study the breeding biology, Elton was to do the parasites, and Ford the protozoology. In addition, A. D. Gardner of the medical faculty agreed to monitor the bacteria, especially the spirochaetes. The accent on parasites was due to the belief that rodent cycles might be caused by the outbreak of epizootics among overcrowded animals.

The small team began their work on British rodents in Bagley Wood, a piece of ancient woodland owned by St John's College, and located only a few miles from Oxford. Trapping was started using twenty household breakback traps purchased from Selfridge's department store in London. The first night's catch was three wood mice (*Apodemus*) and one bank vole (*Clethrionomys*). This disappointed them, but after 600 nights of trapping they came to regard that first catch as rather a good one in relation to effort in the field. This Bagley Wood study, the first systematic work of its kind, was to continue until 1928.

A young departmental technician, Denys Kempson, was assigned to help them for a few months, and became interested in the problems of trap design. Denys developed his remarkable abilities elsewhere for some years, and then returned to work with Elton for the rest of his working life. But at this time, Denys was replaced by a lad sent by J. Armitage, who was known to Elton from the Spitsbergen work and was now a master at Stowe School. The new assistant was A. D. Middleton, who remained with Elton for the next ten years. A couple of years after sending Doug Middleton, Armitage sent another of his laboratory assistants, Richard Ranson. This young man proved to have a remarkable talent for breeding small mammals in cages, and when the fieldwork was wound down, he became John Baker's assistant and looked after the laboratory stocks for three years. Ranson returned to work with Elton and Middleton in 1932.

In 1927, as the Bagley Wood fieldwork was being brought to a close,

Charles Elton setting mouse traps in Bagley Wood, near Oxford, in 1926. Courtesy of Charles Elton.

Charles Elton and E. B. Ford staking out a trap line in Bagley Wood. English gentlemen were accustomed to wear a collar and tie in the field. Courtesy of Charles Elton.

Geoffrey Elton suddenly died at the age of thirty-three. Charles Elton was deeply affected by the loss of his older brother. Through Geoffrey he had been introduced to the beauty to be found in life by observing living things, and introduced to the notion that there must be principles governing their lives. He had already resolved to continue to investigate population fluctuations, especially the epidemiology, and now he had the idea of creating a dynamic memorial to his brother in the form of a special research institute.

A proposal to the Medical Research Council for a grant to study the diseases of wild animals was turned down. But the proposal was sympathetically regarded by the assistant secretary of the MRC, Landsborough Thompson (well known in later years as an ornithologist). Thompson offered to help Elton rewrite his application for a try elsewhere, and gave him a valuable lesson in grantsmanship. First, he admonished him for writing about something as vague as *climate*. "This worried them," he explained. Then he pointed to the word *disease*. "*Never* talk about anything negative like that!" So in every place in the application where

John Baker setting traps in Bagley Wood, as a working member of the Mouse Gesellschaft. Courtesy of Charles Elton.

the word appeared, it was replaced by *health*. Elton now had a proposal to study the health of wild animals. It was sent to the Empire Marketing Board and produced a grant for three years. These things were much simpler in those days. Elton put in this proposal on a Friday, for a substantial amount of money, and received the grant on the following Tuesday, that is, after three working days. In the 1980s it takes longer than that to clear the check at a bank.

Thus Elton and Middleton were enabled to begin the Oxford Rodent Investigation. Doug, or "Dog" as he was usually known, was well versed in country ways and courtesies, as his father had been head gardener to the Sitwell family. This proved to be most valuable in dealing with farmers, foresters, gamekeepers, and petty officials in rural areas. Elton began to gather together the world literature on animal plagues and fluctuations; he had an ambition, for example, to gather into one room everything published about lemmings. His fieldwork was concentrated on *Microtus agrestis*. Middleton had the bright idea of utilizing the newly planted pine forests of the Forestry Commission in the Scottish and Welsh hills. A few years after planting, the young trees were surrounded by dense grass, providing a vole paradise. At the same time John Baker embarked on the first laboratory study of the factors causing breeding seasons in voles (2, 3, 4, 5). Ranson had established laboratory colonies of the field vole (*Microtus agrestis*), the continental form *Microtus arvalis*, the larger Orkney vole (*M. orcadensis*), as well as wood mice (*Apodemus sylvaticus*) and hedgehogs (*Erinaceus europaeus*) (120). Ranson's success in producing tame stocks of wild small mammals was not merely due to persistence and selection; he seemed to have an uncanny rapport with them.

The populations of wood mice and bank voles in Bagley Wood built up to a peak in abundance and then suffered a crash, as predicted. The team failed to find a cause for the rapid decline in numbers of both species. But in the course of 71,768 trap nights and the exhaustive postmortem examination of more than 2,000 specimens, they learned a great deal about what was going on in a population of small rodents, and they developed practical and intellectual techniques for tackling such research. Now that so much work has been carried out on small mammals in a number of countries it is difficult to appreciate how much new ground had to be broken in Bagley Wood by the Oxford Mouse Gesellschaft.

The grant from the Empire Marketing Board ran out in 1931, and by then the Hudson's Bay Company was also feeling the effects of the world depression. George Binney, who had been so supportive of Elton's work, put the firm into a temporarily difficult cash-flow position by moving into the department store business (with great foresight not appreciated at the time) and he was replaced by a nominee of the Bank of England. But in 1930 Elton had been nudged by fate towards his goal by meeting a well-to-do gentleman from Washington D.C., Copley Amory. Amory owned land on the shore of the Gulf of St Lawrence and had become interested in changes in the numbers of animals because of his concern for the people of the region; they suffered great economic hardship in years of scarcity of fish and game. As a keen fisherman, Amory was also upset by the periodic disappearance of the salmon. He decided to organize a conference to which he would invite all of the experts. He began to plan the conference in 1929 and while in England in 1930 he visited the London offices of the Hudson's Bay Company to get their advice. They sent him up to Oxford to have a chat with their consultant in such matters. Amory was sufficiently impressed by Elton to invite him to become secretary of the conference.

The Matamek Conference on Biological Cycles took place in a remote French-speaking village on the Gulf of St Lawrence in 1931, a year in which nearly all of the fluctuations characteristic of Canadian animals hit their lowest point. That year "there were virtually no fur animals, no mice, no game birds, no lobsters, and no cod or mackerel" (48). There must have been a few salmon, however, for Elton recalls Matamek as being the only conference he ever attended at which the chairman of a session walked in and threw two salmon on the table, announcing with pride that he had just caught them. The conference brought together a remarkable and diverse assemblage of people. There were university professors, wildlife managers, trappers, Canadian Indians, and the captain of the Quebec icebreaker. Elton remarked that he would have liked to take the captain back to Oxford, where breaking the ice could be very arduous. Aldo Leopold was there, and R. G. Green. The president of the New York Zoological Society, Madison Grant, had been invited, but he was in poor health and sent Reid Blair as his representative.

If the Matamek conference had not taken place Elton would have persisted in looking for support for his work until he found it. He had been thinking about setting up an independent research institute for some years, but his interest in the subject was what took him to Matamek, not the necessity of raising funds. Nevertheless it was as a result of Amory's conference in 1931 that the Bureau of Animal Population was created in 1932. Reid Blair was so impressed by Elton's contribution that he rec-

ommended to Madison Grant the provision of a grant for Elton's research. The New York Zoological Society provided the funds to operate
a small unit for two years. The name of the new institute was decided
while its founder was lying seasick in his bunk on a Cunard liner returning to Britain. It was chosen by analogy with the Bureau of Biological
Survey in Washington D.C., an organization Elton admired because it
produced fine field-studies by tough, scholarly men. He felt that while
bureau was not terribly British, it was appropriate for a clearinghouse for
information. The use of the singular *population* instead of populations,
which people tended to use in error (and still do) was grammatically
correct. And besides, he says, the unusual usage certainly tended to
make it stick in people's minds.

The proceedings of the Matamek conference were recorded by a
shorthand writer, and the transcript of about 300,000 words was
shipped to Elton, who had the terrible job of cutting it to a reasonable
size for publication (58). The most important outcome of the conference
was, of course, the opportunities it provided for making friends of colleagues previously known only through correspondence or from their
publications. This was to be an important function of the Bureau of Animal Population too. Besides providing a clearinghouse for information
and an atmosphere where people could work with a minimum of distraction, it was the venue of a thirty-five-year international conference.

2 The Early Years: 1932–1938

On January 25, 1932, Charles Elton was able to announce the birth of the Bureau of Animal Population. Professor Goodrich, Linacre Professor of Zoology and Comparative Anatomy, was the kindly midwife who assisted at the birth and also provided a temporary home for the infant. Elton was permitted to keep his room in the department, and was given the use of a cellar for keeping voles. An old shed in the quadrangle was made available to serve as a laboratory. Goodrich could not give any financial support but the housing was guaranteed for three years. It was up to Elton to find funds for staff and expenses.

The room had tables for three: Elton, Middleton, and a graduate assistant, D. H. S. Davis. Inevitably, because of Middleton's nickname, the three occupants were known as Top Dog, Middle Dog, and Bottom Dog. Davis was an Oxford graduate in zoology and forestry. He worked mainly in the field, in England, Scotland, and Wales. A notable contribution to knowledge of voles was his discovery of a short-term activity rhythm in *Microtus agrestis*. (28) The small team was soon expanded by the acquisition of Richard Ranson as the cellar master.

In February 1933, Elton sent a report to all who had helped the Bureau through its first year. This showed that the university contributed £100 of the total £1,624. The rest came from the New York Zoological Society ($2,000 = £564), the Royal Society (£ 500), the Ministry of Agriculture and Fisheries (£ 360), and Imperial Chemical Industries (£ 100). Elton has remarked that if he can be said to write good English, it came from learning Greek at school and from writing grant applications.

This first report contains the Bureau's statement of policy:

> The main aim of the Bureau is to get further knowledge
> of fluctuations in numbers of wild animals, with special
> reference to disease and other factors causing them. We
> have not attained a basis of organisation which enables
> changes in numbers and outbreaks of disease in certain
> species to be studied from year to year, and in some in-
> stances successfully forecast. . . .

The small staff was attempting to cover a lot of ground. Species receiving attention included: Canadian snowshoe rabbit, Canadian lynx, musquash (muskrat), beaver, lemmings, rabbits, squirrels, voles, and wood mice. In addition, Elton was looking into possible links between the diseases of wild and domestic animals, Arctic fox and sledge dogs, voles and sheep (44). In summary, he reported:

> The present method of attack on these problems is (1) field intelligence system, which enables fluctuations in numbers to be followed in a general way, and periodicities worked out; (2) forecasts of periods of abundance and scarcity enabling intensive work to be focused on (3) pathology of diseases, especially epidemics (4) ecology and parasites (5) the control of breeding seasons. (6) At the same time more accurate census methods are developed. The work becomes more accurate and easily transferable to the laboratory as each stage is reached. It has reached different points in different species.

Elton's focus on practical problems brought strong support from the Agricultural Research Council for the permanent establishment of the Bureau within the university. They sent a strong recommendation that the Bureau should receive some form of long-term endowment.

The policy of concentrating on species of economic importance, especially efforts to find ways of forecasting periods of scarcity and abundance, was welcomed by the young Soviet ecologists, whose work was virtually unknown outside of Soviet Russia. Amicable correspondence with some of these isolated workers, especially with A. N. Formozov and N. I. Kalabukhov, led to exchanges of publications. Key papers were translated from the Russian, as guides to the voluminous and often wordy literature. This cooperation was terminated by the 1939–45 war, and the xenophobia of the Stalinist regime prevented renewal when the war was over. Decades later, when an Oxford graduate was attending a conference in the Soviet Union, he was approached by an elderly academician, who asked him if he would carry a small package to Elton. It contained an egg-timer, mounted on a wooden stand which bore the carved likeness of a fox, and when Elton turned it over, he found a neat inscription, "From one old fox to another—N. I. Kalabukhov." From 1932, the Bureau also functioned as the focal point for published British work on population ecology, as it was the physical and spiritual home of the newly created *Journal of Animal Ecology*, for which Elton was the first editor. For the first six volumes, Elton was assisted by Middleton, and for the next dozen by Dennis Chitty. H. C. Gilson was brought into joint editorship for volume 20 (1951), in preparation for taking over full responsibility for volume 21. And he took his editorial prerogatives very seriously, not hesitating to send back papers by Bureau people, and accompanying them with peevish and pedantic comments.

Elton found time to give a series of broadcast talks for amateur naturalists on the BBC. The talks were published in book form in 1933, and present his ideas in the simplest possible way (45).

When the Bureau's second annual report was issued on December 15, 1933, the weak base of financial support was little strengthened, but Elton had been awarded a Leverhulme Fellowship, which he had been advised to apply for by Sir Stephen Tallents, secretary of the Empire Marketing Board. Although Leverhulme grants are for the support of personal research, Elton used most of it to hire a secretary for the Bureau. A full-time research worker, Tom Warwick, had been hired to study the incipient muskrat and coypu problems, arising from escapees from fur farms (143, 144, 145). He spent most of his time in the field, which was just as well, as there was no space for him in the Bureau's room in the Zoology Department. The first secretary-typist was in place, and a part-time employee, Mary Nicholson (Max Nicholson's wife) was going through the Hudson's Bay archives in London. The small staff were still working on a wide range of topics, most of which attracted small grants from interested parties. Some grants were guaranteed for several years; that from the New York Zoological Society would carry on until September 1934, and the Royal Society's until September 1935. Grants from government departments were inevitably annual, but the Department of Scientific and Industrial Research was paying the travel expenses involved in the Hudson's Bay archives research for two years.

There was now a greater emphasis on finding good census methods for "such animals as voles, wood-mice, moles, and such game birds as the partridge." Ranson's success in breeding laboratory stocks, and the availability of good field stocks in the conifer plantations was leading Elton to concentrate his limited resources on vole population dynamics. A conifer plantation provided a good, consistent study area only for a limited time; as the trees grew they created more and more shade until the profusion of grass cover was reduced. Later still, the accumulated needles produced a vole-free habitat, except for patches where trees had died, and along the firebreaks. Most of the plantations were owned by the Forestry Commission, and as voles destroyed some young trees, the commission granted £50 towards the research. As the city of Liverpool planted conifers in its reservoir catchment areas, the city fathers were persuaded to contribute £25. The main manufacturers of safety matches in that era, Messrs. Bryant and May, chipped in another £25. By today's standards of expenditure on research those grants appear minuscule, but when inflation over the decades is taken into account, especially in relation to salaries and gasoline costs, it is seen that they provided significant support. The sum of £100 sterling, had then the purchasing power of at least $5,000 today.

On December 10, 1934, Elton issued the last of the annual reports that were private publications. The Bureau had now completed the third and final year of its trial run. Oxford University had passed a decree making it an official unit of the university, with authority to continue operations for another three years, "with whatever funds it succeeds in collecting." For the next three years there were reports under the university imprint, but Elton was still free to shape them as he wished. Grants available in 1934 were almost the same as in 1932, but Elton had obtained some grants for his personal work and support, and they went into the general account. The Agricultural Research Council was pressing the university to make a commitment of funds, and had succeeded to the extent that the university had included a reference to the needs of the Bureau in its quinquennial report to the University Grants Commissioners. But financial support beyond the end of 1935 was not assured. Elton stated his determination that, whatever the financial problems, the Bureau would in any case carry on by some means or other.

Besides reporting on the North American survey, with maps showing the extent of the network of observers, the 1934 report records the setting up of a network in Britain. Several hundred observers had been recruited to comment regularly on the presence or absence of red and gray squirrels in their localities. From these reports, the spread in Britain of the introduced gray squirrel *Sciurus carolinensis*, was being mapped (105). The squirrel survey was one of Middleton's projects, but his main work was still with game birds, especially the partridge (104). Major Eley, of Imperial Chemical Industries, manufacturers of shotgun cartridges, had set up a game research unit at Knebworth in Hertfordshire, but was still supporting the Bureau's research. The vole *M. agrestis* was still the native species receiving most attention. Besides the regular estimates of numbers by means of standardized trap lines, there were estimates based on the trace-census, a method of estimating vole density arising from their habit of leaving little piles of grass cuttings, as well as fecal pellets, in their runways through the grass. Counting the piles of grass, and the droppings (not in the same places as the cuttings) within sample quadrats, had turned out to be a good means of assessing relative abundance from place to place, and from time to time in the same area. An outside observer would wonder what the fieldworker was up to, scrabbling about in the long grass, and peering into its depths. Back in the cellar beneath the Oxford University Museum, Ranson had succeeded in producing abundant stocks of voles by abandoning the sterile laboratory methods traditional in mouse husbandry, and letting the voles burrow in deep layers of fragrant litter. But he was still having trouble with *Apodemus*.

Elton's ability and industry had made him some good friends inside the university as well as outside it. He enjoyed the support of the registrar, Douglas Veale, to such a degree that he later wrote, "Without his wisdom and guidance through the diabolical complications of University and Government administration, the Bureau would never have got started at all." With help from such friends, the problem of financial support through the critical years 1935 and 1936 was resolved. Elton's Leverhulme Fellowship ran out in July 1935, but he got a Christopher Welch grant to carry him through to November. Several small annual grants were renewed for another year, and most important, the university at last gave both emergency and long-term funds:

> In the autumn of 1935 the University Council set up a Committee to investigate the position and scientific work of the Bureau. It also recommended a temporary block grant of £600 to cover central costs during the period November 1935 to July 1936. As a result of the Committee's favourable report, the Council decided that the Bureau ought to be continued as a University institution, provided funds for central costs, amounting to an irreducible minimum of £ 1140, could be guaranteed for five years.
>
> The sequel was that the University was able to guarantee £850 per annum from the increased Quinquennial grant of the University Grants Commission; Corpus Christi College offered a five-year Senior Research Fellowship to the Director (£300 plus £30 pension contribution); and the Agricultural Research Council guaranteed a sum of £1100 spread over five years (47).

These grants would supply about half of the Bureau's expenses at its current level of activity; Elton would have to find the rest. But the survival of his separate institute seemed to be assured. It would not grow, but then he was not interested in growth. In fact, he began the section of his 1935–36 report dealing with staff matters, "Great importance is attached to having a staff small enough to retain the elasticity and personal contact required in ecological team work." The BAP was just about the size he thought it ought to be. D. H. S. Davis had gone to Sierra Leone to work for the London School of Tropical Medicine. Later he moved to South Africa and became the first ecologist to work on plague in that country. He was replaced by Dennis Chitty, who was the son of a Bristol surgeon but had elected to study forestry at the University of Toronto. This resulted in his acquiring a heavy Canadian accent as well as a Canadian wife, Helen, and no one who met him took him to be English.

Tom Warwick finished his survey of the muskrat and took a post in Edinburgh. In September 1935, at Howard Florey's suggestion, Elton acquired the part-time services of P. H. Leslie, who had been working on whooping-cough bacilli. "George" Leslie had graduated in physiology and had been prevented from completing a medical degree by illness. He began by working on *Salmonella*, which was being considered as a causal factor in rodent cycles, but soon his remarkable flair for mathematics came to be realized, and at age thirty-five he began his career in statistical theory and population dynamics. Another helpful association arranged by Florey was with Dr. H. Q. Wells of the Pathology School, begun in June 1936.

In his *Animal Ecology* and elsewhere (40) Elton had put forward the idea that the fluctuations in numbers of Canadian fur-bearers might be related to the sunspot cycle. In his diffident manner, he now discarded the idea (47). The long historical record which had been assembled from the Hudson's Bay Company's archives ruled out the idea of a correlation with the sunspot cycle, which had a slightly longer period. Decades later, ecologists were still attributing to Elton the belief that the sunspot cycle was the primary cause of the lynx and snowshoe rabbit cycles. What Elton did continue to believe was that a correlation with *some climatic oscillation, interacting with natural interspecific fluctuations of the Lotka-Volterra type, must be postulated to explain the phenomenon.*

The few annual reports which were issued while the Bureau was an independent institute gave its director an opportunity to express his views as he chose. Later, when it was a subdepartment of the university, he became a contributor to a departmental report prepared by someone else. Elton's reports for the fiscal years ending July 31, 1936, and July 31, 1937, contain material which merits being reprinted here. Although written before the first rocket had lifted into the stratosphere from Ger-

many, it anticipates a point of view which became commonplace only
after the earth was first photographed by Americans on the moon.

> . . . the earth is a unique natural park of life, whirling
> through an otherwise lifeless space. In this natural
> park, separating in various proportions from the com-
> mon stream of materials and energy, are over a million
> forms of life, evolved into an unstable and highly com-
> plex network of communities. Of these forms of life,
> man is one of the most powerful; but his power over
> the smaller organisms that cause disease is still quite
> limited. In conflict or competition, but mostly just in
> company with these other organisms, man has slowly
> developed first an awareness, then an interest, and fi-
> nally a scientific approach towards these problems. Na-
> tive knowledge, then natural history, and now ecology.
> Within ecology, animal population research occupies
> an important place.
>
> The practical facets of this science are the fluctuations
> movements and supply of fishes, whales, fur-bearing
> animals, game birds and animals; the conservation of
> wild life for its own safety and man's enjoyment; the
> control of pests (that is our competitors) in trees, farm
> crops, stored products, and water supplies; and, more
> threatening than the rest, the control or avoidance of
> disease in man and his domestic animals. . . .
>
> Population research is impelled by a variety of mo-
> tives. Ecologists seek to assemble first a general descrip-
> tion, and also to some extent a history, or the state of
> animal populations, whose inter-relations and fluctua-
> tions are conspicuous and universal properties of natu-
> ral communities. From a wider point of view they form
> inside the ecosystem of the whole world and surround-
> ing universe, complex capillaries through which flow
> streams of matter and energy, subject to laws that
> are still almost unformulated in physico-chemical
> terms. . . .
>
> There is no human being who is not directly or indi-
> rectly influenced by animal populations, although in-
> tricate chains of connection often obscure the fact. Pop-
> ulation problems are as much part of the fabric of daily
> existence as is the weather. It is quite as interesting to
> know about changes of population as changes in the

weather, and equally important. What is different is that not only do animals have this influence on man, but man has an increasing power over the animal populations that still throng the world. There is less of a moral problem about going out on a doubtful day without an umbrella, than there is about ordering the destruction of a species on the chance that it may be doing harm to human interests.

In the 1936–37 report Elton also made some remarks about the state of ignorance of game populations in Britain:

In spite of the hitches and diversions of the flow of wealth through the countryside, it is plain to the ecologist that research should consider the wealth of farmlands and forests as a whole, and that it is one of his jobs to discover how much of that wealth is being wasted, diverted unwisely from animals to man or from man to animals, and to provide a reliable framework of scientific facts on which a policy of wildlife exploitation or conservation can at any rate be discussed.

It does not seem to be realized how very little we know about what really goes on among the inter-acting populations of wild mammals and game birds that inhabit our cultivated and moor lands. For instance, there are no authentic figures of the reproductive potential, life expectancy, or sex ratio of the common rabbit. There are no statistics about the food of the fox, whose choice of food may hold the balance between hunting and the poultry farming industry. We do not know how much rabbits and hares avoid one another or why. . . .

In the year this was published, a report was presented to the House of Lords by a select committee on Damage by Rabbits. The voluminous report contained more than 270,000 words. Less than 100 words were devoted to the subject of research. Elton planned to make good this lack of knowledge by starting a study of rabbits in 1938: "This research, which will be difficult and will take three years to begin with, may well be expected to produce data of general interest for population theory."

The Bagley Wood study had been wound down, but the voles and mice were not to be left to live out their brief lives (they were now known to be annuals) undisturbed. Francis Evans, a Rhodes Scholar from Haverford College, Pennsylvania, arrived at the Bureau and set to

P. H. Leslie in the library of the Bureau of Animal Population, 1949. Photograph by Denys Kempson. Courtesy of Charles Elton.

work to study the habitat preferences of bank voles and wood mice. Thus it is to an American, Professor Emeritus Francis Evans, of the University of Michigan, that the distinction belongs of being the Bureau's first graduate student, or as they came to be styled, "long-term visitor." Following up Elton's pioneer work with rodent parasites, Evans showed that the number of fleas on a mouse remained about the same, even if they were removed each time it was captured. This constancy proved to be due to the existence of a flea reserve in the nest to which the mouse returned between captures. Francis Evans remained in touch with the Bureau during his long tenure as professor at Ann Arbor, and returned for a sabbatical year in 1962 (61).

George Leslie now began the series of studies which would bring him into prominence in a specialized field of statistics. He began with data on *Microtus agrestis* from two sources; there was Elton's field data, and precise information about the reproduction and mortality of the voles raised by Ranson. Elton had expressed his view that the actuarial methods developed by insurance companies were inadequate for application to small-mammal population analysis. They provided a beginning, but they were really just a description of death rates at different ages. Insurance premiums were not calculated on the basis of reproductive potential at those ages. On the other hand, practically all of the estimates of reproductive increase in animals failed to take natural mortality into account. It should be possible, he felt, to develop a single index which would take into account both births and deaths, in the absence of predators. When this problem was put to George Leslie, he was able to respond by working out *little r*, the *intrinsic rate of increase* for the vole, *Microtus agrestis* (95). This was the first species of mammal, apart from man, for which the requisite data were available. Leslie and Ranson used the term, "true and natural rate of increase," in the same sense it had been used in human demography by Alfred Lotka (100). Later, Leslie was to calculate *little r* for other species (96). At this point he was, as he was accustomed to murmur modestly, "just playing about with the figures."

A significant advance in understanding vole population dynamics was made by Ranson, who found he could count the numbers of young before birth with reasonable accuracy (121). By gently exploring the internal topography of female voles from the outside with the bare fingers, he could count the numbers of developing embryos, and with experience estimate their ages. This provided some unique information about uterine mortality, otherwise obtainable only by killing numbers of females at different stages of pregnancy. The true mice, like *Mus* and *Apodemus* have more muscular abdomens and hold themselves more tensely when handled, but even with them, it is possible to find out whether or not they are advanced in pregnancy.

Ranson's laboratory stocks were yielding reliable information about biological constants to help understand what was happening in the field. But postmortem examinations were still being carried out on large numbers of trapped voles. In 1937, this brought about an unexpected shift of attention; A. Q. Wells discovered a new form of tuberculosis in voles from eleven areas. It was absent from the captive stock. There did not seem to be any correlation between population density and disease; a seasonal fluctuation in its occurrence was probably due to changes in the age structure of the population of short-lived animals (149, 150).

The last of the prewar annual reports (1937–38) seemed to suggest a changed frame of mind in the founder of the BAP. There was no longer a need to stress the idea that the work was of value. The institute had been running productively for five years and there was every prospect of permanence. In his "Review" of the year, his usual forum for explaining

Dr. A. Q. Wells in his study at Shipton Manor, Oxford. Working with the BAP, he discovered vole tuberculosis in 1938. This was later developed for mass protection of school children against human tuberculosis. Courtesy of Charles Elton.

why the work needed to be done, he now explained *how* such work ought to be organized and prosecuted. His remarks are now of the kind we would have read in his next general textbook, if he had chosen to write one. He would like his readers, still mainly supporters and donors, to have a deeper understanding of the Bureau's work. Instead of speaking of censuses in a general way, he now gives eight stages in an animal's life, from the fertilized egg to the old, nonreproducing adult, at which discrimination in counting is relevant to understanding what is going on in the population. He lists nine different causes of mortality, and remarks:

> Although these factors interact with one another, this list comprises the immediate causes of death in a population. "Wear and tear" is the mortality that occurs under optimum conditions, as in laboratory stocks of animals for which the life curve is worked out. In its later stages wear and tear is known as senescence, but it appears to be a fundamental characteristic of organisms that a certain proportion of them fail at every age,

> without death being (at any rate in our present knowl-
> edge) attributable to any of the other basic causes. . . .

Several decades later, the concept of wear and tear was being studied by Dennis Chitty, although it then bore a different label, *stress*. The differing reactions to stress of animals of different ages and born in different conditions of overcrowding was to lead to the concept of differences in genetic quality. As in most categories of modern ecological thought, the complex pattern of the fabric contains at least one colorful thread spun from natural material many years before by Charles Elton.

> Since numbers are the product (and a very complicated
> product) of reproduction and mortality and move-
> ments, it is obvious that the study of all four is required
> in order that a complete explanation of any of them can
> be given. For instance, the potential increase of a spe-
> cies is often calculated on the basis of its reproductive
> powers, without allowance being made for the inci-
> dence of wear and tear mortality. It is as if a motor car
> manufacturer were to calculate the number of cars in
> existence in five years' time without allowing for the
> internal wear and tear which "kills" a great many be-
> fore they reach old age. Equally, a one-sided study of
> mortality may give misleading results. Again, in esti-
> mating the effect of predators on a population of their
> prey, we must constantly ask the question, "what
> would they have died from if they had not been killed
> by a predator?" The pathologist who finds a certain
> death rate in his experimentally inoculated animals,
> needs to know what the normal death rate would be
> from wear and tear, in order to be able to plan his ex-
> periments on a sufficient scale. . . .

The small team suffered a great loss at the end of 1937 when Middleton left to direct the ICI game research station. The parting was amicable, and ICI continued to support the Bureau, as well as joining a liaison committee which was formed to keep the two game-research units working in concert. Elton took the opportunity to state a principle which was to become the Bureau's policy:

> It is realised that one function of a University research
> bureau should be to maintain a supply of well-trained
> investigators, who can carry into the region of applied
> research the scientific methods they have learned.

Middleton had been working on population research
at Oxford for thirteen years, and played a large part in
building up the present organization of the Bureau. In
the early days of the small mammal investigations at
Oxford he assisted in the Bagley Wood survey, making
among other things, a special study of the common
shrew. In 1928 he began to develop the system of co-
operating with the Forestry Commission on trap censuses of voles which proved so profitable in later research on vole populations. During this period he also
created a very large series of contacts in all parts of the
country, from which the most important result was the
grey squirrel survey, continued periodically ever since.
In 1932 he began to turn most of his attention to game
research, and in five years had organized an excellent
system of census and other field work on private es-
tates, which laid the basis for a long-term study of the
partridge, a study that he is continuing in his new post.
Recognition should be given to Middleton's part in developing such a large number of the researches which
the Bureau is doing. His remarkable knowledge of the
shape of the British countryside and the habits of country people and animals has been constantly drawn
upon by his colleagues, with whom he continues to co-
operate closely.

The loss was only temporary: Doug Middleton soon returned to the
Bureau, seconded from ICI, for the duration of the war with Germany.
Meanwhile, his place was filled by H. N. Southern, who was recruited
to work on the rabbit. Elton had trouble getting financial support for
investigating an animal which everyone took for granted:

I approached five agricultural bodies (private and national) for grants. All refused saying that the ecology of
the rabbit was only of zoological interest. After the war,
Harry Thompson, one of our war-time staff who later
became director of a MAFF research laboratory, consulted me about research on rabbits. I suggested that
the most useful practical thing would be to get an un-
equivocal measure of rabbit damage to crops. This was
eventually done and came to figures of about three per
cent. When the British rabbits died of myxomatosis, a
figure of £15,000,000 was worked out for the total sav-

A. D. Middleton gassing rabbit burrows, 1940. The technique was so effective that research on rabbit biology was abandoned for the duration of the war. Photograph by Mick Southern. Courtesy of Charles Elton.

> ing caused by cessation of rabbit damage. It is a curious coincidence that a sum of this order was at the same time knocked off the farm subsidy by the Minister of Agriculture. I think this is quite an interesting bit of social history.

Mick Southern was to remain a member of the small Bureau permanent staff until its demise in 1967. If Elton can be likened to the head and vertebral column of the BAP, three of its limbs were now in place. Marie Gibbs was his right-hand "man," while Dennis Chitty and Mick Southern were his two leg-men. The other limb would be the technical genius, Denys Kempson, who did not join the Bureau until after the war. In 1938, D. K. was in Oxford, working in Julian Huxley's research unit within the Department of Zoology and Comparative Anatomy. He moved with Huxley to King's College, London. King's was evacuated to Bristol during the war, and when the time came for a return to London, D. K. was reluctant to return to the big city. He returned to his native

Oxford, where there was temporary place in the department due to Leonard Small's absence on military service. When Small was demobilized and D. K. was about to be declared redundant, Goodrich arranged for him to join the Bureau as its full-time technician. Then the Bureau of Animal Population was functionally complete.

3 The War Against Waste: 1939–1945

In the spring of 1939 scientific workers throughout Britain were notified that, in the event of war with Germany, they would be exempted from military service in order to carry out research. Elton did not wait to be told what kind of research would be appropriate; he had firm ideas about it. Within his chosen field he was just as much concerned with practical matters as with theoretical principles. In his lectures, he constantly drew examples from species of economic importance; from agriculture, forestry, fisheries, animal husbandry, and human affairs. It was entirely in character, therefore, to take the initiative and offer the services of the Bureau to investigate losses of foodstuffs, both in storage and in the field, caused by vertebrate pests. Such research had bloomed briefly during the 1914–18 world war, and had ceased soon afterwards. He sent off a memorandum to the Agricultural Research Council, pointing out that the Bureau's staff could rapidly change over to research on pests of grain. His people had no direct experience with *Rattus* or *Mus musculus*, but they had unique experience with related rodents.

Early in August 1939, the ARC approved Elton's plan in principle, and three weeks after the outbreak of war the university agreed. The small Bureau staff was expanded by newly funded positions and by the recruitment of some energetic volunteers. Their efforts were to be concentrated on the natural history and control of rodents, and of some less destructive native pests. The main targets were the brown or Norway rat (*R. norvegicus*), common in towns and countryside, the black, ship, or plague rat (*R. rattus*), known in the U.S. as the roof rat, and found in Britain almost exclusively in dock areas, and the ubiquitous house mouse (*Mus musculus*). Work on the European rabbit (*Oryctolagus cuniculus*) had already been started by Southern (130). At various times during the war seventeen individuals were involved. These included James Fisher, Humphrey Hewer, John Perry, Monica Shorten, Richard Freeman, Harry Thompson, and J. S. "Sharon" Watson. And Doug Middleton returned on loan from Imperial Chemical Industries.

Almost all of the Bureau's wartime research has been written up, either in reports to the ARC, in published papers in journals, or in the three-volume work *Control of Rats and Mice* (16). The methodology grew out of the prewar experience, and discoveries about the lives of rodent pests influenced the direction of research when the war was over. The three volumes are surprisingly readable, especially Elton's masterly summary in the first volume, in spite of Chitty's insistence, as the main editor, that everyone toe the line and leave out everything of a personal nature.

The Bureau team, under Elton's low-key but powerful leadership, introduced some novel ideas into British rodent control. First, that the success of control measures could not be measured by a body count; finding some dead bodies did not mean that numbers had been significantly reduced. Second, that a reduction in numbers might encourage breeding or fresh immigration from neighboring populations. Third, and following logically from these two notions, that spasmodic efforts to kill pests when they reach high densities were wasteful and misleading.

It is impossible to recall now how little mammals had been studied in Britain, or anywhere else for that matter before the great surge of intellectual activity which followed the turmoil and pain of the war. But it is even more difficult to comprehend, in view of the vast sales of poisons, that very little was known about the properties of rodenticides. The Bureau staff now had to branch out into applied chemistry as well as applied ecology. Poisons recommended for rodent control included simple inorganic compounds like arsenious oxide and zinc phosphide, some synthesized carbon compounds, and some naturally occurring alkaloids. Even those in common use were neither precisely formulated nor standardized, and they were not applied in a consistent fashion. The only common feature seemed to be that all of them killed some rodents.

The first step was to find out what the available poisons would do under controlled conditions. The roles of the various members of the team can be seen from the authorship of the succession of reports. Freeman wrote up and edited most of the work on poisons, with *Methods of Bioassay* being contributed by Leslie and Thompson. (Harry Thompson took over the laboratory animal stocks when Ranson died in 1944, having worked himself to exhaustion.) The relationship between particle size of arsenious oxide and toxicity was studied by Rzoska, Leslie, and Ranson; and Leslie wrote a special section on the statistics of bioassay. The team also extensively investigated the time-honored rat-poison Red squill, an extract of the plant *Urginea maritima,* with the reputation of killing noxious rodents but not other animals. Its efficacy was apparently due to the inability of rats and mice to throw up, as many other mammals do when they ingest something harmful. After an exhaustive search of the literature, and conducting tests themselves, the Bureau workers concluded that Red squill's reputation was based on ignorance and wishful thinking by salesmen. It was unreliable and generally toxic.

Laboratory tests had to be followed by field trials. One of the first practical problems encountered concerned the provision of containers, or bait stations for the poisons. This was a more exacting problem in wartime than before, as the containers had to be made from materials that were not reserved for military purposes. Bait containers had to be weatherproof, easy to transport and distribute in numbers, impervious to domestic stock, pets, and native animals that were not pests. Ideally they should be made from something cheap, or adapted from something commonly used for other purposes. Two main types of "permanent" bait containers were settled upon, and described by Elton and Ranson. One was an ingenious wooden box which had an entrance tunnel below a raised platform, and made the poison hard to get at by any animal larger than a rat. It could be dismantled for transport. Known as the "P-3," it became standard equipment after the war for local authority "Rodent Operatives" and private enterprise pest-control firms. The second, for rural situations, was made by blocking one end of a standard pottery drainpipe. As these were in common use on farms, they were ready to hand, and it only took a little Portland cement to seal one end. Their main drawback was their weight, but in situations liable to disturbance by livestock that was an advantage. The drainpipe container was easier and cheaper to make than the P-3, but it was less specific to rodents.

In addition to the lack of objective evidence about poisoning, there was little knowledge of the brown rat's behavior. Could a rat remember the food associated with the pains of a sublethal dose of poison? Could a rat discriminate between poisoned and unpoisoned food? In his account of the investigation of the brown rat, Chitty pointed out that the basic problem was that of carrying out a census of the rat population before and after poisoning ("treatment").

For getting an idea of how many rats were present before trying to poison them, the Bureau adopted the apparently simple method of finding out how much plain (that is, unpoisoned) bait was consumed by the rats to whom its attention could be drawn. ". . . census baiting had great advantages, being suited to a wide range of conditions, requiring little specialized training and reducing the subjective element to a mini-

Research station on Giles' Farm, Oxford. Here "new object avoidance" behavior was first observed by Richard Ranson in 1940. Courtesy of Charles Elton.

mum. . . . it is no contribution to substitute a doubtful set of quantitative data for verbal claims, which at least have the advantage of being transparent." Offering a colony of rats a new source of food did not turn out to be simple at all. In attempting to find out why the results were so erratic, the team began to make fundamental discoveries about rat behavior. Their laboratory was an old shack on a pig farm. Their most important discovery from watching rats in the gloom of the piggery shack was that the animals were psychologically disturbed by quite minor changes in their physical environment. They avoided new objects, even new sources of food, or sources of food that were moved about. This was reported by Monica Shorten (Mrs. Vizoso) in volume two of *Control of Rats and Mice:*

> Rats are credited with great cunning, and failures to poison them are attributed to their great powers of recognizing harmful objects. As will be shown . . . brown rats have a natural form of behavior which often enables them to avoid destruction, but which is also displayed with many objects which can have no possible association with danger. . . . this behavior was first en-

countered by Ranson during the design of the P3 . . . since rats seldom entered and took poison bait until after a lag of some days. Soon afterwards Chitty found that rats also showed initial avoidance of harmless objects such as pieces of wood and metal placed in their environment and undertook many of the studies described. . . . (16)

The avoidance of new food had been observed and recorded by Doty, working with rat infestations in Hawaiian sugarcane fields.

From many personal observations it is our belief that, with ample natural food present in a cane field, many rats are suspicious of any newly discovered food supply and only nibble at or entirely refuse to eat it for 2 or 3 days; even though it is of the best non-poisonous material. Thus it is not so much a case of their detecting the poison in the poisoned bait as it is their general suspicion of any new food supply. (30)

As far as we know, the first evidence that anything new is avoided simply because it is there, was recorded in Ranson's notebook one evening in 1940, as he watched rats inside the old piggery on Giles' Farm just outside Oxford:

. . . In the previous ten minutes 39 rat visits to the wheat pile had been observed; but after bread had been placed, although a number of rats entered the shed, they nearly all turned back. Only one rat, in fact, seriously attempted to pass the bread. It first crept up behind and showing great hesitation, attempted to get between the loaf and the wall. After two attempts, which brought it level with the loaf, it gave up and went around to the other side. Here there was plenty of floor space for it to keep well away from the loaf, but after again getting level with it the rat went back and out of the shed.

Now most discoveries are matters of fact, arising from observation rather than intellectual activity, but now and then a number of facts, which may have appeared unrelated to each other, come together in somebody's mind and produce new understanding of a process or of a basic trait. Such flashes of creativity produce knowledge of how and why things happen, and not merely what happens in a particular situation. The recognition of *new object reaction* was an advance in pure science made in the quest for practical ways of killing rats. Awareness of new object reaction in one rodent species led the researchers to look for it in others; after the war, the census methods for voles were affected by it. As in most biological matters, new object reaction turned out to be variable, both within a species and between different species. In 1940 there was no turning aside from the work in hand to investigate the newly recognized phenomenon. It was sufficient to draw attention to it and to take it into account when planning and interpreting control operations.

It was established that new object reaction in the brown rat was elicited by (a) placing new objects near a food supply; (b) using a different container for the food; (c) moving a familiar object to a new position; (d) changing the type of food, (e) changing the illumination, making intermittent noises, or in general disturbing the environment. Removing a disturbance did not, apparently, produce a similar reaction. But that depended, of course, upon how long the new conditions had been maintained; in time the new became the familiar. The erratic results of poisoning attempts were now explicable, so the manner of carrying out poison campaigns was changed. From that time on rats would be offered food to which no poison had been added. Then, when they were feeding confidently, as shown by a plateau in the daily bait "takes," the addition of poison to the same food was much less likely to trigger new object reaction. Rats ate much less of the poisoned food than they had before the poison was added. This was not because it was less palatable, but because they suffered what became labeled "warning symptoms," a euphemism for agony, and stopped feeding altogether.

Harry Thompson investigated poison-bait consumption in various experiments, and Julian Rzoska studied the behavior of rats which had recovered from poisoning. His most important discovery was that rats with experience of poisoned food showed aversion to that same food even if it no longer contained poison. His results could be summarized thus:

(a) An identical poison bait was refused.

(b) A new poison in a base harmfully experienced was refused.

(c) An experienced poison in a new base was mostly accepted.

(d) A new poison in a new base was accepted.

These developments drastically changed the techniques of poisoning rats, and became part of the new dogma after the war: a follow-up treatment always utilized a different poison and a different bait base from that previously used. Before the Bureau team worked on the problem, pest control firms had used the same materials over and over again.

"In 1939 there was practically no government organization of rat control in rural areas. The MAF employed one technical officer whose main duty was to advise on pest destruction, but he had no staff and no authority to carry out rat destruction. A publicity campaign for a special rat week in November of each year was about the sum total of the official rat control."

Thus began Middleton's splendid essay on rat control in rural areas in Britain (16). It was his talent for dealing with country people that enabled him to translate the Bureau's basic research results into practical field procedures capable of execution by dozens of field assistants, many of whom had little experience of working in the field; most were members of the newly formed Women's Land Army. He attacked the problem on a wide front; in farm buildings, corn-ricks (see below), hedge-banks and fields, root-stores ("clamps" of mangolds, potatoes, and carrots) and rural garbage dumps. The method subjected to broad field trials was hole-baiting. Prebaits (bait base without poison) were placed well inside rat holes using a long-handled spoon (at first an ordinary tablespoon tied to a stick) and followed after a few days with poison bait. This was a novel procedure; on one occasion Middleton was accosted by police officers (country bobbies) who decided to take him into custody as they concluded from his strange behavior on a railroad embankment that he must be a saboteur. They decided not to arrest him after he gave them a crash-course in rat control by spooning cyanide salts into holes. Middleton knew that the selection of suitable people was the factor vital for success in pest control:

> It is almost impossible to lay down anything more than general principles about the number and siting of bait points to ensure effective poisoning of an infestation, since this varies so much with the number of rats and the nature of the environment. The success of the work lies mainly in distributing the baits in such a way that every rat has an opportunity of eating sufficient prebait to arouse interest and of taking enough poison bait for a lethal dose, with a minimum of effort and disturbance.

Probably the main cause of waste in rural Britain at that time was the traditional method of storing grain on the stalk. Wheat, oats, and barley were cut, dried in field stooks, and then built into freestanding thatched ricks. These structures, which looked like small thatched cottages without doors, stood in the fields or close to the farm buildings in the "rick yard" throughout the winter. Within these rodent havens there was food, shelter from cold, and cover from enemies. With time the sheaves packed down under their own weight and were difficult for rats to burrow into, so they burrowed underneath, or got up under the thatching. Many a rick had rats in the basement and the attic, and mouse superfamilies in the layers between.

When the grain was needed for market (there were no grain-fed cattle), portable threshing rigs were brought in by contractors and the ricks were dismantled from the roof down. As their homes and food supplies were removed, layer by layer, the rodents escaped into the fields or farm buildings. As the farmers usually took no steps to kill them, except occasionally for amusement, a proportion always survived to reinfest the next generation of ricks. With the disruption of normal commerce by the war, some ricks stood undisturbed until the next harvest. The growth of rodent populations was encouraged and losses of grain, paradoxically, became greater at a time when grain was more valuable.

In order to get an objective assessment of the losses in this special habitat, Elton initiated, early in the war, the collecting of complete rick populations. These thousands of bodies of rats and mice provided material for the study of their population dynamics as well as body counts. At the same time, the Bureau measured how much wild rats of various ages really ate, as distinct from how much they spoiled. With this data it was possible for the first time to give a measure of the food rats removed from the human population. It amounted to about a ship-load a month. Awareness of the scale of the losses, and the results of the field trials reported to the ARC led to the promulgation of "The Rats Order" (1941). Under the provisions of this regulation, the owner of a rick or the threshing contractor was required to erect a fence around the operation, and to kill the rats instead of allowing them to escape. Later on, the order was modified to cover house mice as well.

After the war, these procedures continued to be enforced for about twenty years, by which time farming practices and economics had changed so much that corn-ricks disappeared from the countryside. Consolidation of farms into bigger units, with greater capital investment, eliminated the storage of grain on the stalk. Cereal crops were now threshed and dried and stored in bulk, as had long been the practice in North America and Australia.

The gathering of complete rick populations of rats and mice provided the first substantial material for the statistical studies of George Leslie (97). Although undertaken for practical purposes, it served, as well-designed applied research usually does, to strengthen the intellectual muscles of the Bureau, and affect the direction that future long-term studies would take.

Deep in the ground under every "western" city there lies a network of pipes and tunnels for carrying away human waste and bathwater and the contents of the kitchen sink. These sewer systems provide a home for rats, but the relative prosperity of the rats changes with social changes in the human habitations above. Rats cannot subsist on human feces; they depend upon food being flushed into the sewers instead of being placed in garbage containers. Sewer rats do well in areas which are thickly populated by humans who are neither very rich nor very poor. Such people, especially apartment dwellers, tend to waste food, often flushing it into the drains in useful fragments, instead of converting it into an unavailable sludge with grinding machines installed in their sinks. Nowadays we know quite a lot about the distribution, population dynamics, and behavior of rats in sewers, and how to keep their numbers down at politically acceptable levels of cost. We know also that the stories of rats invading houses from the sewers are untrue; there is practically no interchange of animals between sewer and surface infestations. When London was heavily bombed, sewers were opened and many buildings temporarily abandoned, giving rise to fears in the minds of public health authorities that hordes of rats would move out of the sewers and establish surface infestations. No one knew whether or not these fears were well founded. Sewer workers knew that rats were normally to be found throughout the system, as a disease carried by rats, Weill's disease, was an occupational hazard on the job. The sewer-rat problem was tackled by Humphrey Hewer in the Oxford sewers, and handed over to Chitty and Freeman when Hewer was appointed no. 2 rodent officer in the Ministry of Food. Humphrey later took over as chief rodent officer and made a fine job of applying the Bureau's discoveries to practical rat control, for the duration of the war. John Perry, a graduate from the University College of North Wales, also became involved during his work with Watson in the London docks area, and after the war worked with ministry officials conducting field trials in sewers. Perry compiled the chapter "Control of Rats in Sewers" from reports by all involved.

Baiting rats on the surface was arduous enough. Baiting them in sew-

Dennis Chitty taking some fresh air while baiting rats in sewers. The King's Arms tavern nearby has caught his attention. Courtesy of Charles Elton.

ers involved difficulties of a higher (or lower!) order. Sewers are very interesting places to visit, as I can personally attest, but they have some fundamental disadvantages as places in which to study highly mobile small mammals. Modern sewer systems separate storm water from domestic and industrial effluents, but the systems in the older sections of British cities were not so sophisticated. In those old single systems the flow could change from a trickle to full bore in a few minutes following a downpour. This washed away the bait stations, and put the lives of the researchers at risk through drowning; there always had to be an assistant standing by at an open manhole to watch the weather. There was also the potential hazard of leptospiral jaundice, a nasty disease carried in rat urine, even when greatly diluted. This had been known to get into the bloodstream through a superficial scratch.

In terms of rat biology, the main findings from this laborious work underground were that (1) the distribution of rats in sewers was correlated with that of the humans in the buildings above, (2) that the avoidance of new objects was just as important in the baiting of sewer rats as it was with surface-dwelling rats, and (3) that the black rat (*Rattus rat-*

tus), bearer of the fleas that carried bubonic plague (the Black Death of the Middle Ages), was not to be found in sewers. Even in the London docks, where black rats were present in greater numbers than brown rats, all of the sewer rats were of the latter species.

The methods of poisoning sewer rats devised and tested by Chitty and his colleagues formed the basic procedures of all British local-authority rodent operatives from 1945 until the late 1950s. Similar techniques had been introduced in the United States by the Bureau's counterparts working out of Johns Hopkins University (59, 60). But then a revolution in rodent pest control was brought about by the application of anticoagulant chemicals. In 1942, the causative agent of a disease of livestock known as sweet clover disease had been isolated by Dr. K. P. Link in Wisconsin. It was a compound which prevented normal clotting of blood in minute intestinal scratches and brought about death by internal bleeding. After testing the substance on rabbits, Link made the brilliant suggestion that it "might make a good exterminating agent." The substance was synthesized and named Compound 42, or WARF 42 (after the funding agency, the Wisconsin Alumni Research Foundation). Further work was delayed by the war, but the rodenticidal potentiality was recognized in 1949. In June 1950, compound 42 was registered for general use and the trade name Warfarin was adopted. The first field trial with rodents was apparently by J. A. O'Connor, who suggested that the technique would remove the shyness caused by previously experienced sublethal doses of acute poisons (114). The new method led to the development of numerous other anticoagulants, and for a time it seemed that the costs of rodent control had been reduced for all time. Surely, it was argued, the manner of bringing about death could not lead to the development of immune breeds, as observed with DDT and insects. But before two more decades had passed, prebaiting and poisoning with acute poisons had been reintroduced in areas where both rats and mice were found to be immune to anticoagulants. This fascinating story cannot be expanded upon here, but much of the research was carried out in the Infestation Control Division of the Ministry of Agriculture and Fisheries, by staff trained by BAP "graduates."

The rat problems of the Port of London Health Authority were not limited to those arising from the infestations of ships and the risk of importing bubonic plague, or from persistent infestations of both species in dockside warehouses (brown rats downstairs, and black rats, known as roof rats in the U.S. upstairs). There were peculiar industrial problems arising out of wartime factory conditions. Many factories producing supplies for the armed services had been established in old buildings in the dockland area. In one of these, the rats' nests in the walls were heavily infested with mites. The mites were crawling about the work areas and attaching themselves to factory workers, mainly female, causing serious disruptions in production. This odd example of biological interference in the war effort led Dr. Morgan to ask Elton to put a research team into the docks to study the whole gamut of rat problems. Elton sent in James Fisher, assisted by Sharon Watson. Both were Oxford graduates, but as different as the proverbial chalk and cheese. Fisher was a bright self-promoter and front man, while Watson was a quiet, industrious worker. After a time, Morgan pointed out to Elton that Watson was doing all of the hard work. The Boss had a quiet talk with Fisher over a cup of tea and gently but plainly suggested that he transfer to some other kind of work more suited to his considerable talents, such as scientific journalism. Fisher became a well-known writer and broadcaster, and editor of Collins' New Naturalist series. He also wrote a number of scholarly books, mainly about birds. In this instance, as in several others, Elton's toughness when it was necessary to fire someone led to diversion into a successful career. Watson's modest manner made him well able to establish good relationships with the dockers (longshoremen) whose cooperation was essential, just as Middleton had won the friendship and respect of farm workers. Sharon also visited other important wartime ports such as Liverpool to study how rats affected the efficient operation of the ports, and to train local people in control methods.

It became clear that although the black rat had been displaced in the countryside and most urban habitats by the brown rat since the middle of the eighteenth century, the black rat was still the dominant species in dock areas. As the Bureau researchers had begun their work with the most common rat, and made headway in reducing its numbers, they attempted to apply their knowledge and techniques to this closely related species. They soon found that the methods used for dealing with the brown rat could not be transferred and produce the same results. The black rat behaved, in some ways, like a big form of house mouse; it was more curious about new objects than afraid of approaching them, and more erratic in its visits to baits left in place for a number of days. Nevertheless, for logistical reasons it proved to be necessary to use a prebaiting procedure similar to that used for brown rats. As labor costs were an important consideration, the development of dovetailing schedules for work crews was one of the most important aspects of rodent control that the Bureau people had to work on. In the system that was worked out, one control crew was used to deal with two separate infestations at a time. On the first working day of the week the crew placed token baits without poison in one location. These were renewed on the third day and replaced by poisoned baits on the fifth day. The same crew placed token baits in a different location on their second and fourth days. Poison baits were placed in the second location on the first day of the next week. The following day the bait trays were picked up from the first location, and next day the bait trays were retrieved from the second location. The remainder of this second week was spent scouting for new infestations. The people employed in rat control in ports were mainly involved in fumigating ships, which they knew well from stem to stern. They all had ship master's certificates, and Elton found them "a very interesting crew indeed."

While most of the Bureau's wartime research was focused on rats, Elton recognized, right at the start, the importance of the house mouse as a spoiler of human food. As early as June 1940, he was personally surveying mouse problems in Oxford grocery shops. When Middleton's success in fumigating rabbits temporarily lowered their importance as pests requiring study, Elton asked Mick Southern to take over the house mouse, so Mick became the Bureau's main mouse man in 1942. In 1943, he was assisted by Mary Laurie in looking at reproductive rates of mice living in different habitats, and this work became her subject for a postgraduate degree. Then, in September of 1943 Julian Rzoska began to devote most of his time to helping with the vexing mouse problem. Julian had escaped from Poland when it was overrun by the Nazis in 1939, but had been interned in Romania. With some fellow officers he had escaped from there and made his way, via Italy and France, to Britain. The Agricultural Research Council had offered his services to Elton, whom he had met just before the start of the war. Although a hydrobiologist by training, Julian embraced the rodent work with enthusiasm.

Between November 1941 and February 1942 Elton had been examining the serious problem of mice in strategic stocks held in Ministry of Food buffer depots. British agriculture was not then sufficiently developed to provide the population with bread. In the years that Britain stood alone, there was only two weeks' supply of flour and grain in storage at any one time. The supply was held in stores near centers of population, and many of the buildings were unsuitable, with easy access for rodents. Within the buffer depots grain was stored in hessian (burlap) sacks which were piled in large stacks. Each stack was a mouse paradise, providing superabundant food, shelter, and nesting material. When the mice nibbled at the sacks, whether for the grain or the fibers, the grain bled out, causing the stacks to slump, even to collapse. Some of the spilled grain could be rebagged, but much had to be diverted from human use to animal food. Even if used for animals, rebagged grain could be hazardous if poison bait from control attempts was mixed with it. The standard prebaiting and poisoning methods developed for the brown rat didn't work with house mice. Within the mouse paradise of a buffer depot they would sample new sources of food, but the amount taken from any one food source was minute. This provided many opportunities for the ingestion of a sublethal dose, which, as with rats, caused an animal to stop feeding altogether until it recovered. House mice seemed to be quick to explore new things in their environment, even traps, but when a population was heavily trapped, there was some evidence of trap avoidance or shyness. (A decade later, my own work suggested that this might be due to inherited behavioral differences between individual mice) (26).

Southern set about studying the mouse as a wild animal, for although the same species had been bred in laboratories by the thousand for experimental work, almost nothing was known about the wild form. At the same time, he began field trials to test every tentative fact as it came to light. Julian Rzoska, with his experience in poison trials with rats, applied himself to poisoning mice and found it much more difficult. It was he who suggested fumigating the stacks of bagged grain with carbon dioxide, by enclosing the stacks in tarpaulins with "dry ice." This technique was later applied to ricks by Middleton. One of Southern's most significant discoveries was that the mouse had an extremely small *home range*. This meant that a very large number of baiting points had to be used to ensure that some were within reach of every mouse. A poison campaign was a failure if it left a few mice alive. If only one pair survived, or a solitary pregnant female, the very rapid rate of reproduction in the absence of predation quickly built up the population again. Volume 3 of *The Control of Rats and Mice* was almost entirely written by Mick Southern, and it became the instruction manual for training the rodent operatives who had to continue the war against mice in buffer depots for many years after the war with Germany was over. For the Cold War with Russia caused Britain to maintain strategic stocks of grain and flour for another fifteen years.

The research of the small group in the Bureau of Animal Population, and of Emlen's small group in Baltimore, Maryland, greatly added to knowledge, but the main contribution to pest control made by both groups was not in the gathering of facts; it was in the changing of attitudes. Before the war, rodent control was carried out by a handful of "professional" rat catchers, who preserved much of the medieval belief in secret, exotic remedies. They were amateurs, even when they were not charlatans, in that they kept what little they learned from experience to themselves and their apprentices. After the war, the professional staff employed in pest control, whether they worked for the government or private enterprise, carried on the wartime professionalism, both in research and in applying it to practical problems. The greatest change was in recognizing the importance of the human element. Sometimes, changing the behavior of the workmen in a factory or store could reduce a pest problem more than tackling the pests themselves. Rats were shown to be more easily controlled if piles of rubbish and debris which provided cover were tidied up. If workmen took away the scraps left from lunch instead of leaving them in the workplace, rodents would leave a factory instead of multiplying. There could even be a significant reduction in the reproduction of mice in a flour store if the leaking faucet in the washroom was fixed. These were the practical attitudes carried over into the wartime research by the Bureau team, and inherited by their successors in the Infestation Control Division of the Ministry of Agriculture, Fisheries and Food.

The considerable delay in getting *Control of Rats and Mice* into print proved to be advantageous. It permitted the inclusion of valuable review material missing from the original reports to the ARC, and of a wider range of references. Richard Freeman had acted as librarian to the research team, and he developed a valuable bibliography. The Clarendon Press, being the more conservative arm of the Oxford University Press, would consider publishing the work only if a cash subsidy was found, as they thought it too specialized to sell. Elton obtained the necessary cash from "an anonymous donor," and when sales exceeded expectation, he was in the happy position of being able to offer to return it. The donor refused, and it became a useful contingency fund for Bureau staff travel. Anyone acquainted with British rodent research would require only a single guess to identify the donor; Sir John Ellerman was the only person alive who had the necessary interest and affluence.

Some of the Bureau's work on agricultural pests other than rats and mice was continued during the war years. Southern carried on his classic study of a marked population of rabbits in a fenced warren, but moved on to the house mouse when it became clear that the rabbit was being kept down by arbitrary methods developed by Middleton. Fisher worked

John Perry and Sharon Watson during a tour of duty for the British Foreign Office with staff of the Sudan Public Health Department, 1945. Courtesy of Charles Elton.

on rooks for a time, and the survey of squirrels was maintained. But the main war effort was on rats and mice.

When the war ended in 1945 and supernumerary members of the Bureau's staff returned to their prewar jobs or vocations, Richard Freeman went back to University College and Pat Venables to his farm in Wales. Humphrey Hewer went back to Imperial College, from which he had been seconded, and soon began to study seal populations in the Farne Islands, a project the Boss likes to attribute to exposure to the Bureau way of thinking. Monica Shorten (Mrs. Alberto Vizoso), after taking a few days, at Elton's suggestion, to "look at squirrels" made their study her professional life's occupation (125–27). Watson, with backing from Professor P. A. Buxton, got a Colonial Office grant to study rodent problems in Cyprus (including black rats living up pine trees) (146). Later he went to work in wildlife research in New Zealand. Julian Rzoska became the University of Khartoum's first professor of zoology. Harry V. Thompson carried the Bureau's cool style into the Ministry of Agriculture and Fisheries, where he mastered the machinery of administration and built

When Sharon Watson emigrated to New Zealand to carry out research in exotic pests, he sent back this photograph to demonstrate the immensity of the problem. Courtesy of Charles Elton.

up a research group which was able to carry out field studies with a minimum of bureaucratic interference.

The Bureau continued to act as a research and advisory team for the ARC until the end of July 1947. By then the MAFF had set up the Infestation Control Division which, besides taking over the research function, had executive responsibilities for funding and inspecting the rodent control operations of local authorities. Infestation Control Division also dealt with practical aspects of the control of insect pests and the fumigation of buffer depots. It was without doubt the best organized and administered unit of its kind. But as agricultural practices and storage methods changed, and government spending in other areas changed budgetary priorities, it was gradually dismantled by successive governments. Its destruction took about forty years to accomplish and is irrelevant to this story.

The small Bureau team remaining in Oxford was unchanged, except for the grievous loss of Ranson. But their quarters within the zoology department were cramped, especially now that much of the fieldwork was closed down. University life was getting back to normal, with its

Richard Ranson at work in Southern's experimental rabbit warren at Sheep-
stead; setting baits to test taste-preference of rabbits. Photograph by H. N.
Southern. Courtesy of Charles Elton.

petty politics, and with the postwar resurgence of academic and teaching activity there were territorial troubles ahead. Then Elton had a tele-
phone call from the secretary of the University Chest; if a plan could be
produced in time for a meeting to be held at 10 o'clock the next morning, the Bureau might be able to expand into a wartime hospital which
had been built in the grounds of an Oxford College. Elton worked all
night and produced a diagram in time. The university administrators
had no problem in accepting his explanation of how he would use the
many rooms he said he needed, including a large one vaguely labeled
"Experiments." But they were rather puzzled by his allocating an office
to "Squirrels."

In 1947, the Bureau of Animal Population and the Edward Grey Institute for Field Ornithology were administratively separated from the Department of Zoology and Comparative Anatomy, and, while retaining
their separate identities, were combined to form a new Department of
Zoological Field Studies. The professor of the old department became
also chairman of the new department. But the Bureau and the EGI now
had spacious quarters in separate wings of a sprawling armed-forces

hospital in the grounds of St. Hugh's College. Now there was space for the staff to each have a room, for a few graduate students, a library, and for a workshop and darkroom for Denys Kempson, a human cornucopia from whom would flow ideas and ingenious inventions.

Both Dennis Chitty and Mick Southern had to spend much of their time after the move working up the wartime material for publication. But each had a major research project planned which would keep him occupied for the next decade at least, and would make an intellectual background for their work with visiting students. Dennis, rejoined in his work by his wife, Helen, now that their children were of school age, returned to the vole work and greatly expanded its scope and influence. Mick embarked on a monumental study of the relationships between the tawny owl and its prey in Wytham Great Wood, part of an estate a few miles from Oxford. The Boss was finishing off "odd jobs" like the study of Arctic fox movements, and developing his ideas about an ecological survey of Wytham Estate.

Professor Goodrich's successor was Alister Hardy, a gentleman entirely sympathetic to the notion that good people should be helped to do whatever they wanted to do. In February 1962, with Hardy's retirement imminent, Elton remarked: "We shall never cease to be grateful to our Professor, Sir Alister Hardy, for saving the situation then—and making conditions for peaceful and uninterrupted and fertile work during the following fifteen years. He has given us the maximum support with the minimum of interference."

Let us now look at the Bureau's happy life during those fifteen years.

4 In a College Garden: 1947–1951

The heart of the City of Oxford is the historic crossroads known as Carfax (pronounced by local residents as "Carfarks"), a name probably derived from the French "Quartre voies." The street running due north from Carfax is Cornmarket ("The Corn"), which expands into a short but grand tree-lined boulevard called St. Giles. There is an interesting mixture of important structures at its proximal end; Balliol College, the Randolph Hotel, the Ashmolean Museum, the Martyrs Memorial, and the public conveniences. At the northern end the ancient church of St. Giles and its graveyard occupy a triangle formed when "The Giler" bifurcates into two main arteries, the Banbury Road and the Woodstock Road. These two meander apart and are linked by a succession of streets which become only slightly longer as one travels to North Oxford. Between two of these, the Canterbury Road and St. Margaret's, lie the lovely grounds of St. Hugh's, one of Oxford's few women's colleges.

The principal of St. Hugh's may, with justification, have regarded the decision by His Majesty's Government to build an ugly squat sprawling octopus of hospital hutments within the grounds as an act of administrative rape. But like so many other structures built in the 1940s, the hospital was only "temporary," and in wartime sacrifices had to be made with good grace. She was less forgiving when, at the end of the war, the University Chest solved a number of problems arising within its growing empire by utilizing the hospital to house three of its odd colonies: the Lectureship in the Design and Analysis of Scientific Experiment, the Edward Grey Institute for Field Ornithology, and the Bureau of Animal Population.

North Oxford was full of vast brick houses with lovely front gardens. Most of these were built for the first generation of Oxford dons permitted to marry while in the service of the colleges. The senior men had the larger houses fronting on the Banbury and Woodstock Roads, and the lesser lights had smaller versions, some conjoined and some in terraces, on the linking rungs. After the Second World War the traffic was still punctuated by stately ladies mounted on tall black Raleigh bicycles with dress guards; the widows and sisters of dons sacrificed in the First World War. Their houses provided rooms for students upstairs, while they retained an apartment at garden level, and their frugal lives were enlivened by shopping expeditions to North Parade or Little Clarendon Street for minute quantities of high-class comestibles.

Spring in Britain was a revelation to anyone, like myself, raised in a land dominated by evergreens, and although the Oxford gardens had been neglected during the recent war, there was still evidence of the original planning for a succession of colored leaves and blossoms. In March and April there were massive outbursts of chestnuts, flowering cherries, forsythias, lilacs and laburnums, and mutant beeches and plums. Many of the shrubs and ornamental fruit trees produced double blooms, collectively a "foam of blossom." Heavy scented Buddleias attracted tortoiseshell and peacock butterflies, and at night, dodging the low-hanging branches on the narrow sidewalks one walked into heavy clouds of fragrance, which added another dimension to the already formidable culture shock. The gardens were dull after July when the original owners normally went abroad, and the cycle of beauty was at its nadir.

In addition to the advantage of beautiful surroundings, the hutments provided a spacious, quiet workplace. The only reminder that we were within the college grounds was an occasion tinkle of laughter or conversation heard through the high windows. No doubt we were more intrusive, as we worked at all hours. Compared with the cramped quarters in the old University Museum these were spacious facilities. There were now more than twenty rooms of various sizes, comprising 6,000 square feet in total, and the corridors, designed for wheeling beds through, were wide and without changes of levels. The Bureau's portion of the complex was shaped like a tuning fork; two long prongs and a short junction which connected them with the rest of the building. The allocation of space to various functions and people was logical and economical. When one got to know the Boss and D. K., one soon realized that the most apparently inconsequential detail was either the result of profound consideration or of conscious compromise between the ideal and the practicable.

The Field Store, for traps and other paraphernalia to be taken to Wytham, was immediately inside the main entrance. D. K.'s workshop was its neighbor and then D. K.'s store. The right prong housed Marie Gibbs, H. N. Southern, and the large library. In the left wing, the large room corresponding with the library was called "The Museum," and it was the repository for the specimens and data being accumulated in the course of the Wytham Ecological Survey. The Boss had an office adjoining the Museum, and the Chittys' and George Leslie's rooms were also on that side. There were inefficient aspects; Elton was a long walk from the library, but that meant also a long way from the office, something which would be to his liking. Most of the space available for visitors and students was in the one side, while their animal accommodation was in the other. But they were expected to spend much of their time in the field anyway.

On April 18, 1947, Elton made one of his rare speeches, to students, colleagues, and friends gathered in the library, which also served as a social center, and commented: "For sixteen years we have studied rodents, and like them have occupied various partly connected burrow systems that latterly became overcrowded. . . . We must hope that the new paint and great comfort will not in any way corrupt our research abilities." This was not an entirely whimsical remark. Elton preferred to work in a monastic type of environment, without the distraction of frills. He also preferred to keep administration and "bureaucratic bumf" to a minimum. There was no formality, and the clerical control of vehicles, stationery, and equipment had not yet been imposed by the university. I think that the few people who had the good fortune to work in the Bureau while it was housed in the St. Hugh's huts would agree that these were the golden years of their professional lives.

The first postgraduate student at the Bureau after the war was, like the very first of all, a Rhodes scholar. This was Peter Larkin, the 1946 scholar from Saskatchewan, Canada. He had declared his desire to work with Elton when he applied for the scholarship: "The reason behind that was that in 1944 I was one of a field expedition to Great Slave Lake in the North West Territories. One of the older fellows in the party was Dr. J. G. Oughton from the Royal Ontario Museum. Oughton was a very fine naturalist and he told us once that when he went on field work he only took along three books for company—the Bible, Darwin's Voyage of the Beagle and Elton's Animal Ecology. That was the summer I read both Darwin and Elton." "When I arrived in Oxford Charles asked me what I wished to do for a doctorate. I suggested further work on glacial relict Crustacea (*Pontoporeia* and *Mysis*), which had been my Master's topic. After a fairly rigorous cross examination, Charles said he thought I already knew enough about that and suggested two other topics: (1) the ecology of tree-hole aquatic communities, i.e. all the things in stagnant ponds in tree holes; or (2) the ecology of mole populations." One reason

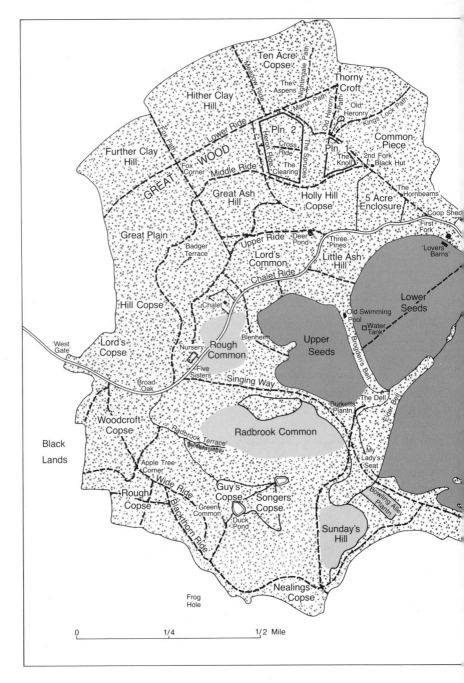

Sketch map of Wytham Estate, with the main subdivisions and landmarks, as designated in 1950. Modified from standard blank form used for fieldwork in the BAP.

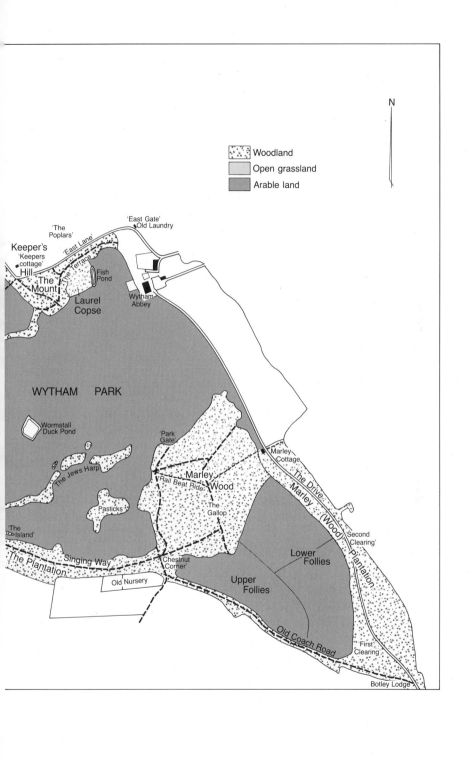

N

Woodland
Open grassland
Arable land

'The Poplars'
'East Gate' Old Laundry

Keeper's Hill
'Keepers cottage'
'East Lane'
'The Terrace'
Fish Pond
The Mount
Laurel Copse
Wytham Abbey

WYTHAM PARK

Wormstall Duck Pond

'Park Gate'
Marley Cottage

The Jews Harp
Marley Wood
Rail Beat Ride
The Gallop
Pasticks

The Drive
Marley (Wood) Plantation

'The Island'
Singing Way
'Chestnut Corner'
'Second Clearing'
Lower Follies

The Plantation
Old Nursery
Upper Follies

Old Coach Road

'First' Clearing

Botley Lodge

he chose the moles was that the project was supported by the Ministry of Agriculture, and that made a car and gas ration available. Elton arranged for him to collect regular samples of moles from the lovely grounds of Blenheim Palace at Woodstock, where "Capability" Brown had done some of his finest landscape planning. Thus Peter Larkin became, for two years, honorary mole-catcher to His Grace the Duke of Marlborough, who used to gaze down on him in wonder from his seat on a very high horse. The young Larkin is still remembered as a clever and amusing man who took on the considerable physical work involved in mole trapping with enthusiasm. Marie Gibbs recalls him holding high and rattling a bunch of traps and exclaiming with glee: "Those moles are really going to get it today!" When he had been working in the field for a few months, Elton asked if he might accompany him to Blenheim. On the way he asked, "How many traps have you set?" and Pete answered, "Twenty-four." "And how many moles will you catch?" "Oh, eighteen or nineteen." When the traps had been picked up there were nineteen moles, and Elton commented, "Good. You've passed the acid test. You can always find out if a fellow knows the animal he's working on, by seeing if he can predict how many he will catch." (Elton recalls offering him water voles, and that it was just as well they did not appeal, as they were all but wiped out by the hard winter of 1947. It was for that reason he steered me off them when I showed interest in studying them in 1949.)

Apart from Leo Harrison Matthews' study of its reproduction, the mole had received little attention since being an object of interest (along with almost every other living thing in the countryside) of Victorian and Edwardian country gentlemen.

Much of Larkin's work corroborated the views of these amateur naturalists, and put them on a quantitative basis. In addition, he studied changes in mole numbers, and the factors affecting their distribution and population density. The regular samples taken in Blenheim Park, and some from another locality, besides enabling him to follow changes in body weight and reproductive condition, provided skulls from which tooth measurements were taken. The degree of tooth wear served as an index of an individual's age. Thus he could work out the age-structure of the population in different times of the year. He found that established colonies of moles were made up of equal proportions of young-of-the-year and older individuals. In newly colonized land, young-of-the-year comprised up to 90 percent of the population.

The onset of breeding in March and April was associated with an increase in the size of home ranges, especially those of males. Home range was the subject of one of Larkin's excursions into theoretical ecology, such intellectual effort being expected of all candidates for the Doctor of Philosophy degree. He defined home range in general terms: an area of a particular shape and size "which is repeatedly traversed and influenced by the animal." There have been many definitions since then, but Peter Larkin's simple concept remains one of the best. Using Larkin's measurements of home ranges and trapping results, Dennis Chitty attempted to estimate "the true density and mean range of movement of moles" and this exercise is appended to the thesis.

The mole (*Talpa europaea*) is an especially suitable species for the study of home range. It spends very little time outside of its subterranean burrows, which thus provide, when excavated, an accurate map of its habitual pathways. In freshly seeded arable land, before crop growth obscures the surface, it is possible to watch the invasion of moles and to measure the rate at which they extend their burrow systems, something impossible with most mammals. But when a number of home ranges are contiguous, as in mature pasture, it is difficult to determine how much trespassing goes on, and how much home ranges are held in common through lack of synchrony in activity patterns. It was this fascinating situation which led Gillian Godfrey to apply to moles the radioactive tracking technique she developed for studying the behavior of *Microtus* (below).

Peter Larkin submitted his thesis in 1948 and returned to Canada. His two-year sojourn at the Bureau was a great success from all points of view, except that the thesis was not published because he became involved in other research. But he made the material freely available to two later Bureau graduates when they decided to write a book about moles (67).

In 1949, four postgraduate students came to the Bureau, a higher intake than in subsequent years. They were a "mixed bag" and arrived in diverse ways, but all eventually successfully completed doctoral theses. Peter Crowcroft was the first to arrive, docking at Southampton in February, fresh (some thought excessively so) from the University of Tasmania, on an overseas fellowship from the Commonwealth Scientific and Industrial Research Organization (then known as the Council for Scientific and Industrial Research). He had been recommended for the fellowship by his employer, Professor V. V. Hickman on the strength of his research on fish parasites, but the CSIR was undecided about what he should study or where. As these fellowships were intended for training prospective scientific officers for its various divisions, the areas of study depended upon current and imminent vacancies in the organization. Eventually, he received a letter telling him he could choose either to continue with his studies of fish parasites, at "some place to be decided," or to study animal ecology at Oxford University, with a view to joining Francis Ratcliffe's new Wildlife Survey Section.

There were some interesting people for a Trematode man to work with, and some excellent laboratories in which to work; Harold Manter, improbably enough, was in Lincoln, Nebraska, Marie Labour in Liverpool, England, and Satyu Yamaguti in Japan. But like any other of his generation of Australians he would have chosen to go to Oxford even if it meant switching to the study of Ancient Sanskrit. Oxford was even more attractive than the great marine research institute at Naples, which Hickman spoke of with awe, Acceptance of this alternative produced the information that Francis Ratcliffe was an Oxford graduate, and a friend of one Charles Elton, who would make a place in his Bureau of Animal Population. This led Crowcroft to go to the small departmental library and read for the first time a slender volume entitled *Animal Ecology*. Ratcliffe visited Tasmania to consult with his staff there and to catch some trout and to look over his new boy. He spent half an hour chatting, still embracing the fly rod he had carried on the plane, said he would write to Elton and set it up, and took off for the Tasmanian Lake District. After some months Crowcroft wrote to Elton saying he had arranged to arrive in February of the next year. Elton had not heard from Ratcliffe, but he found the student's letter "very business-like" and agreed he should come as planned.

Two graduates from the Oxford Department of Zoology and Comparative Anatomy entered the Bureau in 1949. One of these, John Clarke, was the 1946 Rhodes scholar for the state of Western Australia. He came to Oxford in 1947 and completed the work for a B.A. in two years. He discovered Elton's course through contact with Francis Huxley, with whom he went to Gambia in 1948.

He was, he recalls, "totally captivated . . . by Charles' mild whimsical, dry humoured, soft-voiced style. . . . I also enjoyed enormously the lantern slides with which he invariably illustrated his lectures—photographs of Charles trapping animals in some outlandish place, or somebody struggling across some ice making an ecological survey. . . . It appealed very much, perhaps, to my Australian background where I had spent a good deal of my youth wandering through the scrub and bush with my geologist father."

The author picking up mouse traps in Great Wood, Wytham, in 1950. Photograph by Denys Kempson. Courtesy of Charles Elton.

John had met Dennis Chitty earlier. "One of my first memories of Oxford when I arrived in 1947 . . . is of a research talk in the Zoology Department by a forgotten person on a forgotten subject. Sitting next to me on the bench was an attractive figure of a man in a blue duffle coat. He looked at me, maybe recognising from my bewildered innocent-looking face, that I was a new boy, slid (at some personal risk—if you slid along this bench you got splinters in your bottom) along the bench, put out his hand and said, 'I'm Dennis Chitty.'" This made an indelible impression on John as it was the first time someone had troubled to speak to him without an introduction. They kept in slight contact during the next two years, "a tenuous contact," but it seemed natural when John went to the Bureau that Elton should direct him to Dennis as a supervisor.

The other Oxford graduate was Gillian Godfrey, an extremely shy but fiercely dedicated zoologist, who had modest but adequate private financial resources to maintain her in Oxford as a graduate student. By the Easter term preceding her final examinations she had decided she wanted to be an ecologist ("despite Elton's abominable lectures") and wrote to Elton asking if she could carry out some menial tasks in the

John Clarke prepared for hockey at St. John's College, Oxford, 1949. Photograph courtesy of John Clarke.

Bureau during the following term. This led to her spending a couple of afternoons a week during her last term as an undergraduate dissecting moles for Peter Larkin. This experience converted her to small-mammal ecology and to her deciding to turn down the offer of a place at the Plymouth Marine Research Station.

Professor John Clarke prepared for seminar at Durham University, 1960. Photograph courtesy of John Clarke.

Her interview with Elton was awkward. He told her he didn't care to have women in the Bureau just yet. She offered to work as a bottle washer and that did the trick. There wasn't much future in bottle washing he retorted, so she had better come and do research. Dennis Chitty wanted several aspects of vole biology investigated, so Gillian joined the small vole team and in due course became the first woman graduate to complete a D. Phil. within the BAP.

That appeared to complete the complement for 1949. But in October a Fulbright scholar from Colorado, Richard S. Miller, strolled into the Bureau and announced to Elton that he had come to do research. The 1949 batch of Fulbrights had been chosen late and Rick had barely had time to catch the boat to Britain. On arrival at his college his tutor advised him that if he wanted to do ecology he had better go and see Mr. Elton. So he walked around to see him at once. The Boss liked Rick and took him in, suggesting that he pick up a thread of the pioneer Bagley Wood mouse work. Thus Rick Miller began a study of the food preferences and other habits of wood mice and bank voles. The preliminary outline of the research included a note provided by the Boss that summarized the conclusions eventually reached.

John and Gillian set out to get information about *Microtus* in order to test some of Chitty's ideas about population cycles. When Chitty had analyzed his data from the Lake Vyrnwy population that had been studied continuously from 1936 through 1939, he found himself unable to explain their crash in terms of the classic factors: food, weather, predation, and disease. He came to suppose that overcrowding and its associated social strife might have caused the high mortality among young voles born when numbers were highest. Adverse effects of overcrowding could also explain the reduced productivity of females subjected to it. That hypothesis would not have caused any raised eyebrows, even in the 1940s. The Lake Vyrnwy population continued to decline, however, in the next generation, when there was no longer an overcrowded vole society, and when food appeared to be abundant. Chitty felt obliged to offer, as the simplest possible explanation, the existence of an inherited disability: "During the time of their almost complete disappearance in 1938 or 1939 voles were not subjected to any known environmental conditions likely to have caused excessive mortality. The hypothesis is therefore advanced that death was primarily due to adverse conditions to which the parents were subjected in the previous breeding season."

That suggestion not only caused the eyebrows of authorities in population matters to go up, it also raised the hairs on the backs of their necks. This smacked of the Lamarckian heresy! Chitty could not get the paper published in the journal of his choice. But Sir Alister Hardy read it objectively, and communicated it to the Royal Society for publication in their *Transactions*, a most prestigious place.

When Dennis guided John and Gillian into their studies of vole physiology and behavior, the term *stress* was not in everyday use. But it soon would be, for Hans Selye's general adaptation syndrome, or the *stress syndrome* was about to be accepted as the explanation of population regulation in small mammals, as well as the underlying cause of many human ailments and maladjustments. A few more years would have to pass before this notion ceased to dazzle the eyes of population ecologists, and the dynamic nature of genetic inheritance in local populations would dominate their subject. Then the 'Chitty hypothesis" was to become much more respectable.

Gillian Godfrey set out to study "Factors affecting the survival, movements, and intraspecific relations during early life in populations of small mammals with particular reference to the vole." This was rather a large canvas to tackle. The first objective was to persuade voles to nest in small wooden boxes, which could be inspected periodically to get information about litters. With great single-mindedness, and almost no experience with hand tools, she set about mass producing nest boxes in D.K.'s workshop. After his initial consternation, not because of the possibility of her injuring herself, but because he feared she might damage his tools, he diplomatically suggested that she could work undisturbed in the field store. For many days the corridor echoed with the sounds of saw and hammer, and Gillian emerged with large numbers of wooden nest boxes with removable lids.

Her plan was to emulate the fine work of Howard with American small mammals (73). Unfortunately, the British rodent *Microtus* refused to behave like its American cousin *Peromyscus,* and continued to make all of its nests in secret places. Gillian decided to find their nests, even if it meant hand-searching every inch of the habitat. This she proceeded to do, spending days on her knees dissecting every tussock of the long grass on a study area called "The Dell." She was very thorough, found lots of old nests, and learned a good deal about the nesting habits of *Microtus* in that habitat. But she got a good dressing down by the Boss for causing such physical devastation to one of Wytham's few pure stands of *Brachypodium pinnatum*. The damage proved to be superficial and the grass quickly returned to its former lush appearance.

Then, in an inspired change of tactics brought about by the failure of

Gillian Godfrey tracking the movements of a vole wearing a leg-ring labeled with radioactive cobalt. Rough Common, Wytham, 1950. Photograph by Denys Kempson. Courtesy of Charles Elton.

the voles to use her nest boxes, she set out to trace their movements and find their nests by putting radioactive rings on their legs and finding them with a Geiger-Müller counter. She was greatly assisted on the technical side by a physicist with amorous ambitions which were fruitless and ill-conceived. Cobalt 60 wire was obtained from Harwell before the Boss got wind of the project. He was pretty upset by her initiative, but saw that the technique had such great possibilities that she got away with it. This was the first time small mammals had been tracked in this fashion, and it was a breakthrough in that it meant trapping data no longer had to be relied upon to estimate home range in *Microtus* (65, 66).

D. K. was intrigued by the Geiger counter and borrowed it for a few days to test everything in the place for radioactivity, including the luminous dial of Elton's wristwatch. This gave such an impressively rapid series of clicks that it could not be worn again until D. K. had painstakingly scraped off all of the luminous paint.

The half-life of the radioactive cobalt was about five years, so after Gillian had left the Bureau, D. K. was at a loss what to do with the lead pot and the unused wire. It hung in the bicycle shed for some time and then disappeared. Years later, Monte Lloyd discovered it in a hole in a tree in the Pasticks, a suitably isolated copse in the middle of an arable field on Wytham Hill. By then its radioactivity would have been innocuous. But I still have some mental discomfort when I recall cutting up the wire for Gillian with two pairs of pliers, and rescuing bits that flew off by using the screaming Geiger counter.

John Clarke's thesis was entitled, "The response and behaviour of animals at different population densities, with special reference to the vole *Microtus agrestis*," and he set out to monitor events in two confined populations founded at different densities. The notion was that overcrowding would cause a sudden decline in numbers (a so-called "crash"), and a study of when and how the two populations declined would cast light on the factors affecting numbers in free-living voles.

At the same time, John began to observe the aggressive behavior of voles, by staging encounters in pens and in glass tubes similar in size to vole runs. This was a basic description and analysis, with observations on displacement activities associated with fighting. It was a good time to study behavior at Oxford, for Niko Tinbergen, its first lecturer in Ethology, was available for discussion and advice. Tinbergen was exceptionally helpful as an adviser for students outside of his own "school," and I made good use of the opportunity to consult him.

One vole population was established in an old swimming pool in Wytham, discovered while John and Dennis Chitty were wandering about in late September 1949, discussing what direction the work might take. It seemed to be custom-made for what they had in mind. The second population was started in an old water tank which they found in a piece of beech wood only about 250 yards away. This one became known as the "Snake-pit," a name that has stuck to it ever since. There were no snakes in it, but John was a fan of Olivia de Havilland, and at that time she was appearing in a film called *The Snake Pit*. It seemed appropriate to transfer the name of an overcrowded psychiatric ward to the overcrowded vole institution.

The experimental colonies were not to grow and crash as planned. John arrived at the swimming pool one day to check the inhabitants, and found an extra one—a weasel was cruising about under the cover provided for the voles. John seized a piece of asbestos fiber sheeting used as a nest-box cover and raised it high. At that moment, Farmer Wise, who managed the Wytham farm, looked into the swimming pool and cried, "Don't you touch that bloody animal." This was an expression of a regard for natural predators not commonly expressed by men on the land, but John ignored it and destroyed the weasel with a blow to the neck. "That rid me of my passion," he recalls, "but hardly compensated for the slaughtering of part of that population by the weasel."

The same calamity happened in the Snake Pit some months later. On that occasion all 64 of the voles present were killed, and the predator escaped John's wrath, presumably by climbing up one of the poles holding the wire netting on top. There was a positive result, however, for these disasters forced John to sit down and analyse the data he had accumulated over eighteen months of hard work. He found that the curtailed experiment had been worthwhile. "There were important, significant differences between the populations, and within each population with time . . . consistent with some of the detail in Dennis' theory. There were reductions in fertility of the successive generations of voles correlated with the increase in size of the populations, and changes in time in the expectation of life of adults and juveniles." George Leslie provided "enormous help with the analysis of these data," and the work was published in the *Proceedings of the Royal Society* (20).

A third part of John Clarke's research, not planned at the start, was on the effects of fighting stress on voles. This arose indirectly out of a visit by Dr. Wes Whitten, who was visiting Britain from Canberra, looking at animal houses. Whitten came to Oxford to talk with Howard Florey, who was considering a move to Canberra, and he mentioned to John that Hans Selye was giving a talk. John went to Selye's lecture, and was so stimulated by the ideas and the showmanship that he decided to try some simple experiments with voles. A number were subjected daily for three weeks to an hour of aggressive encounters. The adrenals, thymus, and spleen from these "stressed" animals were then compared with those from similar animals which had been living alone (19).

Rick Miller's research, in contrast with that of Gillian Godfrey and John Clarke, in which drastic changes of direction were necessary for completing satisfactory theses, proceeded in an orderly progression, as planned, to a preconceived conclusion. This did not mean less hard work or hard thinking, but it did mean less agonizing. His thesis title was "Activity patterns in small mammals with special reference to their use of natural resources."

A section of the animal room was separated from the rest by a wooden frame and black plastic sheeting, to provide a nocturnal house in which day and night could be reversed and their duration controlled by a time clock. D. K. had modified a number of clockwork drum recorders; each electric impulse now produced a fine prick with a needle in the paper instead of a pen mark. Mick Southern had used them originally, and I also used them for recording shrew activity patterns (107).

In addition, Rick studied the feeding preferences of his rodent subjects, by examination of stomach contents, and by feeding tests (106). But the more subtle and challenging aspect of the study concerned their habitat preferences. Discussions on this topic with Elton provided intellectual contact the rest of us did not have. The Boss and Rick developed an excellent rapport, and when his thesis was complete, Rick stayed on to work on systems of classifying habitats and communities, funded by a grant from the Nature Conservancy, and became coauthor with Elton when the work was written up (56). I can remember being unreasonably envious at the time.

In September of 1950, as the 1949 batch of students were approaching the end of their allocated time at the Bureau, a more mature student arrived with a two-year grant from the Nature Conservancy. Elton had sat with Cyril Diver for a week, interviewing six people a day. Most of the applicants for grants were bright young people from Oxford and Cambridge. But there was one older applicant with a third-class honors degree from a provincial university who had managed to reach the personal interview stage because of his unusual record. This was Eric Duffey, who had spent two years flying seaplanes with convoys, and the often boring duty had led him to study sea birds. Then after the end of the war he had been a member of an expedition to Bear Island. His quiet and controlled manner appealed to the Boss, but would the kind of degree be a problem in this competition? Cyril Diver, as director of the Conservancy had the final word. He saw the gleam in Elton's eye and reassured him, "Don't you worry Charles. You'll get 'im."

When Elton advised Eric to choose a group with interesting behavior, he chose spiders: "An ecological study of spider (Aranea) communities in limestone grassland." The fieldwork was mainly in two of Wytham's grassland areas, "The Bowling Alley" and "Sunday's Hill." Besides ingeniously catching air-borne spiders on sticky canes, Eric used pit traps and quadrat searches. He rapidly became an expert in their taxonomy and breeding biology, and the effects of the vegetation. He had to apply for a London University Ph.D., but the Boss wrote to me: "It was a completely Oxford job, and one of the best theses ever written there."

There were several long-term visitors to the huts in St. Hugh's who were not working for higher degrees. In January 1950, Dr. Christian Overgaard-Nielsen came from Copenhagen. He was a highly trained soil

Denys Kempson, Richard Miller, and Charles Elton in a relaxed moment while surveying vegetation in Wytham Estate, 1950. Courtesy of Charles Elton.

ecologist and he collaborated with Amyan Mafadyen, who had been appointed to the Bureau staff in the previous year, on a survey of the invertebrates in the soil of the Dell. Overgaard later carried out a pioneering study of the energy balance sheet of *Glomeris marginata,* the millipede, which was very abundant there. He used the silica in the *Brachypodium* as a tracer, as people would later use isotopes. The millipede consumed more *Brachypodium* litter than the plants produced each year. This was possible because it ate the same material over and over again. According to Amyan, this was the first indication of the phenomenon known as "microbe stripping," a kind of external microbial digestion now known to be commonplace. Overgaard returned to Denmark in 1953 and became head of the Mols Laboratory in Jutland. He was a very quiet little man, described by Elton as unflappable and of massive intellectual grasp.

In January 1950, also, Winifred Phillips came from Aberystwyth in Wales, where she had been working on rabbit control with a grant from the Universities Federation for Animal Welfare, which was testing alternatives to the leg-hold trap. Winnie needed to use the Bureau library in working up her results. She also gave great assistance that summer to Mick Southern with the Wytham Great Wood small-mammal trapping. For a time she was joined by another UFAW worker, Marie ("Nooky")

Winifred Phillips, a graduate of the University College of North Wales, helping with small-mammal trapping in Great Wood, Wytham, June 1950. Photograph by Denys Kempson. Courtesy of Charles Elton.

Stevens, who later went to work for Harry Thompson in Infestation Control Division.

The Bureau's first "Kiwi" arrived in November of 1950, nervously tapping at the doors of the BAP and the EGI. This was Gordon Williams, on leave from the New Zealand Wildlife Service. An admirer from afar, he had written to David Lack and Charles Elton asking if he might spend time in Oxford, in return for whatever assistance he could give in any capacity. Both had sent him slightly encouraging replies, Elton's slightly warmer than Lack's. As he was in such awe of both of them he arrived in Oxford in a very humble frame of mind, but determined at least to meet these two ecological aristocrats.

When he arrived at the St Hugh's huts, he turned (by chance, he said) to the left into the EGI. But David Lack was overseas, so he proceeded to the BAP to ask if Elton was available. The Boss was in, and he let Gordon pour out his aspirations. After hearing Gordon's recital of his lack of background training in ecology, but his earnest desire to learn, and how he had saved up his leave to give him an opportunity to spend time in

Oxford, Elton suggested that he help Mick Southern with the analysis of owl pellets for a couple of weeks, to see whether he wanted to stay longer in the Bureau or not. Gordon realized that this was a tactful way of saying what was obviously the exact opposite. So Gordon took over the job of mashing up owl pellets with the "milk-shake" machine I had built from D.K.'s store of bits and pieces, and he spent many hours, at all times of the day and night, reading in the Bureau library.

After a week, the Boss asked me what I thought about Gordon, and did I think he was a suitable person to be given space for six months. I was flattered to be asked for my opinion, which he may have wanted because he put Gordon and me into the same category. I thought for a bit and gave him a straight answer, which I recall exactly, "I don't think Gordon will do anything very original, or set the world on fire, but he's an honest, hard-working fellow, and if you can find a job that requires a lot of hard work and not a lot of originality, I think he'll be OK." Elton looked pleased with my answer, and said he agreed. A few days later Elton consulted with Dennis Chitty, and the outcome was that Dennis turned over to Gordon an enormous mass of data the two Chittys had collected on game-bird fluctuations in Canada and the northern United States. Gordon worked it up into a paper for the *Journal of Animal Ecology* (152). After his return to New Zealand, his Oxford experience and his subsequent studies enabled Gordon to move into university teaching, where he set up the first animal ecology course in New Zealand.

Every researcher has to be inventive to some extent. When investigating a problem for the first time, it will either be necessary to modify the techniques which others have used in similar work or to devise entirely new ones. It is normal for most of the techniques evolved during a piece of research to perish at its conclusion, simply because they are not relevant to anything else. When people have become carried away with clever technology, they may publish a paper on it. When they do, it may have to serve as a substitute for the results they ought to have "worked up" and published instead. But now and then, techniques are developed which can be used for a variety of projects, and which are not entirely dependent upon personal skill. When that happens, there has been a step forward.

The husbandry methods for voles developed by Ranson before the war became standard practice in the Bureau's vole room. New students accustomed to the sterile, bare cages in mouse rooms elsewhere, found it odd, at first, to keep animals in cages with several inches of debris in the

bottom, in which the voles burrowed and prospered and, most important, bred prolifically. As voles are much less odoriferous than house
mice, these unhygienic cages were not unpleasantly smelly. It was not
as simple to teach Ranson's palpation technique, as that depended very
much upon personal skill. Gillian Godfrey and John Clarke tried their
hands at this during their first trapping exercises in Wytham. Gillian has
related how John opened his huge Westralian hand to release his vole,
to find it had quite given up the ghost. I can recall finding Gillian in the
vole room, hands streaming blood and face streaked with tears, bravely
pressing on with vole examinations, and explaining with great embar-
rassment, "Oh, but they bite *so* hard!" A few weeks later she was deftly
holding them by the loose skin of the back with one hand, palpating the
abdomen with the other.

The most useful inheritance from the prewar work was in the live-
trapping and marking of mice and voles. Since that historic moment in
1936 when Dennis Chitty released voles bearing metal leg rings in Bagley Wood, there had been little technical progress (10). Corresponding
work in the United States had led to the use of metal ear tags, which had
certain advantages but tended to get ripped out of the thin pinnae of
mouse ears. When Denys Kempson joined the Bureau and applied his
mind to producing a better mouse trap, he and Chitty evolved a trap
which would profoundly change the pace of development of British
mammalogy. Before he worked with rats, Chitty had been preoccupied
with the notion that a vole would not enter a trap which was not open
at both ends, a true tunnel. Now with the knowledge that new-object
avoidance had been the problem, that false lead was abandoned, and
the Longworth trap was born. The design was fixed, and the manufac-
turing operations were carefully worked out by D. K. Then Kempson,
Chitty, and Southern, with Larkin's help, produced the first gross of traps
in 1948. By the time we students arrived in 1949, an arrangement had
been entered into with the Longworth Instrument Company in Abingdon, Berkshire, for commercial production and marketing of the trap.
One condition of the agreement was that the Chitty-Kempson trap
should bear the Longworth name (14). Over the next two decades the
Longworth became one of the two small-mammal traps in general use
throughout the world. The other was, of course, the lightweight folding
trap devised by Professor Sherman.

The Bureau trap consisted of two parts, the tunnel trap and a clip-on
nesting box. The tunnel opened for cleaning, and the treadle spring was

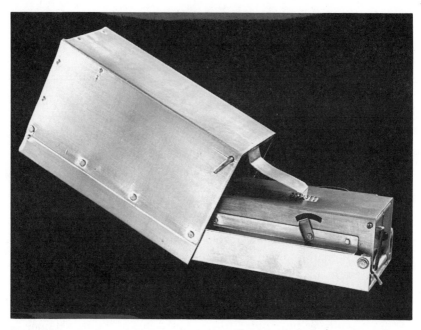

The Longworth small-mammal trap developed by Denys Kempson and Dennis Chitty. Minor improvements were added later but the basic design is unchanged today. Photograph by Denys Kempson. Courtesy of the Ecological Society.

adjustable. It slid inside the nest box for transport. The description was published in the *Ecology* and in the same year the first batch of mass-produced traps arrived from the Longworth Instrument Company's plant at Abingdon. Chitty used to take off for Wales with them and Southern used them in his monumental study of the prey of the tawny owl in Wytham Great Wood. When I began trapping on my own account, having embarked on a study of shrews, there was hardly a day when I did not set some of the traps in Wytham as I explored a variety of habitats. The occasional presence of sprung but empty traps which smelled of shrew led to the realization that pygmy shrews (*Sorex minutus*) were escaping through an opening in the trigger cowling, left to facilitate cleaning. I devised a cover for the opening and this made the Longworths effective for pygmy shrews trapping (22). After some years the Longworth people changed the design and made the addition unnecessary. Thus in a minute way I was able to make a late contribution to the perfection of the design. D. K. was never satisfied with the traps, and he experimented with new designs, seeking to come closer to his ideal trap which would "weigh nothing, fold up to occupy no space, and cost next to nothing to make." Ken Marsland, his assistant and successor, and now chief technician of the Oxford Zoology Department, still has some mysterious plates of sheet metal, like pieces of a jigsaw puzzle which cannot be completed because some pieces are missing.

5 In a Botanic Garden: 1952–1967

The city of Oxford does not have a Zoological Garden, but since 1621 there has been a small, choice Botanic Garden. In its early years it was primarily a collection of plants associated with the study of *materia medica,* but it has a long-standing association with classical plant taxonomy and is now part of the Department of Botany of the university. This tranquil Botanic Garden, a few yards from the horrific traffic of the High Street, is one of Oxford's best-kept secrets. The city does not need to promote it as a destination for the tourist trade; Oxford has more than enough of those in its colleges. Carolus Linneaus, founder of the binomial system of naming living things visited the Botanic Garden in 1736 and showed its curator, Dr. Dillenius, his proposed revolutionary classification system. The good Doctor was so impressed with the potential importance of Linnaeus's work that he offered to share his house and his salary with Linnaeus until death, if the Swedish genius would move to Oxford and complete his studies there. Linnaeus did not take up this extraordinary offer, but he later expressed his thanks: *In Anglia est qui genera curat vel intelligat praeterquam Dillenius,* and he named a Genus of evergreen Indian trees *Dillenia* in his honor. There is a portrait of the chubby and generous curator in the stairwell of the Linnean Society of London, in Burlington House, Piccadilly.

Anyone who has been engaged in taxonomic research cannot fail to feel some emotion when walking on the same gravel paths that Linnaeus trod on his visit to Oxford. For a biologist, the Botanic Garden has the kind of significance the lovely college quadrangles have for students of literature or history. I had an inkling of how Dillenius felt, strolling there with Arthur Cain, then curator of the Pitt-Rivers Museum. Arthur had the unnerving habit of giving the full scientific names of animals mentioned in conversation, that is, after rattling off the genus and species, he kept on and gave the name of the original describer and the date of the publication. This is pedantically correct in biological literature but unheard of in conversation. It came so naturally to Arthur that I cannot think it was one-upmanship. But it was very impressive, especially

The wing of the old Botanic Garden building occupied by the BAP, 1952–67. Linnaeus walked here. Photograph by Charles Elton.

when he quoted Linneaus in Latin and asked innocently, "Don't you agree?"

The Bureau of Animal Population was moved from the huts in St. Hugh's gardens to the old Botanic Garden building on January 1, 1952. Everything was arranged according to well-laid plans, executed under the watchful eyes of D. K. The Bureau occupied approximately one-half of the two-storied building, and the Edward Grey the other half. They were sealed off from one another at ground level, but a corridor ran the length of the building upstairs. A door separated the two institutions, and it was kept locked. The heart of the Bureau, the library, was now in a more pleasant room, with French doors facing the Botanic Garden, but it was otherwise unchanged in ambience. There were the same shelves and filing cabinets, and people still worked and took afternoon tea at the plain wooden tables. Marjorie Nicholls was still unobtrusively in charge, always helpful but never fussing over the frequent visitors who came from all parts of the world to use its unique collection for a few hours, days, or months. For someone frustrated by working for years without access to scientific journals, working in this specialized library was mental bliss. One could pull a reprint from a writer's file, note in its bibliography five or six papers that should also be consulted, and instead of writing off to libraries overseas for photocopies, walk a few steps to other files and pull out every one of them. If in these papers there were others to be looked at, the odds were high that they too were immediately available. Working with reprints and photocopies has physical as well as psychological advantages. The heavy, bound volumes of periodicals, so dear to library administrators, are unwieldy to handle and impossible to keep opened at the right places without recourse to heavy weights (usually other volumes) and this generates guilty feelings about breaking the spines.

If one had to look further afield than ecological journals, Oxford was one of the best places in the world to be; the main library was one of the finest in existence, and there was a special repository for scientific material, the Radcliffe Science Library, adjacent to the Zoo Department on Parks Road, less than half a mile away.

The library was the social center of the Bureau as well as its main research tool. Except at tea-time, when everyone gathered there, the number of people using the library at any one time was so small that the one function never interfered with the other.

At the start of Michaelmas Term, 1952, a new trio of postgraduate workers arrived; Michael Davies, John LeGay Brereton, and David Jenkins. In selecting them Elton once more demonstrated his faith in heterogeneity. John Brereton was another ex-pilot, a bit exhausted by the war, the scion of an Australian pastoral family, with a fine American wife. David Jenkins was a Cambridge graduate, with the coolness and self-assurance one finds in people from "the other place." Mike Davies, on the other hand, was a quiet sensitive graduate from Imperial College.

Brereton undertook "A study of factors controlling the population of some terrestrial Isopods." He concentrated on the distribution and movements of four Isopods that were common in Wytham; *Trichoniscus pusillus* which inhabits litter and dead wood, *Philoscia muscorum*, which lives in litter, mainly in the deeper layers, *Oniscus asellus*, which feeds on dead wood and is the one commonly seen in the woodpile, and *Porcellio scaber*, which climbs tree trunks at night and is found under cover near them. John watched the vertical migrations of *Porcellio* at night, but he did not carry out much fieldwork (6). He was a thoughtful, even contemplative kind of worker. He would put a few woodlice in a dish for a few hours, observe them, and ponder what he had seen for two weeks. I had the pleasure of carrying out some simple experiments with him, to test the palatability of his Isopods according to my shrews (22). Unlike

John Brereton (rolled sleeves) and his mentor, Mick Southern, prepare for a research talk in the BAP library, 1952. Photograph by Bruce Falls.

many Australians he was soft-spoken and gentle in his dealings with others. Whatever I told him about my work, he looked very grave and said slowly, "Jeez Pete—that's vair-y interesting." From October to December, 1952, "Cocky" Brereton shared his room with George Dunnet, a canny Scot from Aberdeen. George had attended the BAP field ecology course in 1949, and he was now being exposed to fieldwork on vertebrates with Mick Southern prior to joining Francis Ratcliffe in the Wildlife Survey Section. He spent much of his time writing up his Ph.D. work on starlings. His supervisor for that work, Robert Carrick, was also recruited by CSIRO.

Mike Davies also worked on invertebrates: "The ecology of small predatory beetles, with special reference to their competitive relations." He studied ground beetles, of which four or five species with apparently similar habits are often found in the same places (27). Competition between similar species, with apparently similar needs was then, and still is for some ecologists, of great theoretical interest. It was a difficult undertaking, more suitable for unhurried long-term research than for producing a thesis in a set time. Elton was not impressed with the result, but it passed the examiners. The thesis begins with a historical account

Members of the 1950 field ecology course taking a break in Wytham. From left, Richard Miller, Miss S. Chapman, Charles Elton, M. Wallace, and Mike Davies.

of Wytham, over twenty pages of it, quite irrelevant to the subject of the thesis. Davies was not able to demonstrate ecological differences between the species, but he conceived a pleasing mental picture of their changes in numbers: "a picture of a thin film or layer of population, with occasional peaks and troughs of higher or lower density, that stretches as far as the physical limits of the organic woodland or grassland community of which it forms a part." The variations in numbers with the passage of the seasons he likened to the changes in the level of the young Thames which, "flowing from its source in the Cotswolds far to the West, meets Wytham Hill standing to the North, embraces it, running around it on three sides before breaking through . . . into the southern valley of Oxford." While the level of the river changed with the seasons, at the end of the cycle it would be just about what it was at the beginning. So it was with the beetle populations. After a year of ups and downs they were as thin upon the ground as they were the year before. But the processes at work remained mysterious.

David Jenkins studied the partridge, with Dennis Chitty as his supervisor. Arrangements were made for him to work on a square mile of

David Jenkins pondering partridge population problems, 1952. Photograph by Bruce Falls.

agricultural land in Hampshire, part of Arthur Rank's estate. This was an excellent place for fieldwork, but David inevitably spent most of his time away from the Bureau. As a fashionable game bird in Britain, the partridge had already received a good deal of notice, and Middleton had studied it in the Bureau's early years (104). But now it was studied in greater depth, with a lot of individually marked birds. Much good information was obtained about seasonal changes in numbers and about the bird's dependence upon certain types of cover (74). In addition to studying population dynamics, David observed partridge behavior, and found correlations between the two. A sudden drop in numbers in January and February was due to aggressive behavior; there was much fighting and chasing, and the poorer or less determined fighters were driven out of the best places (75). This came just before pair formation, so the peak of pairing behavior corresponded with the lowest level of population. Jenkins concluded that the partridges of the 640 acres should not be thought of as one population but as a number of small local populations of constant density at breeding time. Thus the level of partridge production is to a great extent controlled by the amount of suitable cover and the innate behavior of the birds. Shooting is not really controlling numbers, it is a process which removes most of the birds which could not be accommodated by the available cover when January comes again.

The international nature of the Bureau was further shown by the arrival late in 1952 of two more overseas graduates. In October, J. S. Tener, a staff member of the Canadian Wildlife Service, came on a year's leave, and helped with the Orkney vole *little r* experiments. Maxim Todorovic came from Yugoslavia in November, to train for two years. Elton had always placed a high value on maintaining foreign contacts, especially with Russian ecologists tackling practical wildlife problems and seeking, at the same time, understanding of the processes at work. But the Cold War made visits by Eastern Bloc ecologists impossible. Now, thanks to Marshall Tito's efforts to do business with the West, the British Council had funds for grants to students from his country.

Maxim worked with Mick Southern, filling the niche I had occupied for the first six months of 1949, and conducted some trapping experiments of his own in Marley Wood. Before coming to Oxford, he had investigated mole populations, using homemade traps which he carried to and from his study area on his back (141). So while the Bureau's facilities might seem frugal to visitors from North America, to someone from the other side of the looking glass they were luxurious. Maxim was accustomed to work without access to manufactured goods taken for granted as normal store items. He had planned, for example, to monitor the activity pattern of moles with a recorder, but the idea had to be abandoned because there were no flashlight batteries with which to power a field recorder. For Maxim, as for his Tasmanian predecessor, D. K.'s storeroom crammed with war-surplus electric motors, gear-trains, solenoids, relays, a potpourri of nuts, bolts, and screws in various sizes and alloys, and intricate pieces of aluminum and brass (and even bronze) from bombsights and other instruments of destruction, was a treasure house like the cave of Ali Baba. And D. K. had been able to buy all of that marvelous stuff through the university for a few cents a pound.

Rick Miller completed the qualifications for his doctorate in the minimum time and was admitted to the degree in 1951. Both Gillian Godfrey and John Clarke received their degrees two years later. Gillian had to rewrite sections of her thesis, especially those relevant to Chitty's hypothesis, which Lack considered heretical, while John was delayed by being offered a lectureship in Oxford and sensibly taking it. I did not complete my thesis until 1954, having worked on it in spare time while carrying out research on the house mouse for the MAF Infestation Control Division, and commuting every weekend from London to maintain legal status as a resident graduate student.

The year 1953 was a good one to visit the Bureau and stay for tea. Brereton was still there, and Jenkins sometimes. Maxim's command of English had vastly improved, and there was a striking and talented graduate from Richard Freeman's department, Brenda MacPherson, helping Elton with the Wytham Biological Survey. A visitor from the University of Toronto, Dr. Bruce Falls, recorded the informal but earnest nature of a Bureau afternoon tea session. Bruce had the nerve, or lack of nerves, to come into the library an hour earlier when Elton was working at one end of the long table and I was working at the other, to stand beside me, call softly, "Mr. Elton?" and to take a flashlight photograph of the Boss as he looked up. The victim was coldly furious for an instant, as he hates to be photographed under any circumstances. He quickly recovered, uttered a strangled "Good God!" and returned to his work with a sad smile. I thought at the time that he had looked like a startled stork, and was sorry not to have the picture. Then, after thirty years, when circumstances placed me in Toronto, with an honorary post in the university there, I became a colleague of Bruce Falls. One day, at tea in that department, I recalled his audacity and asked him if he still had the negative. He kindly, or unkindly, dug it out of his files and gave me a print. I treasure it.

In the fall of 1953 the Bureau acquired its most talented mature student, Edward W. Fager, and also an Oxford graduate bearing the illustrious name of Thomas Huxley. Bill Fager was arranging his things in his room upstairs during one of my frequent visits. I put my head in his door to say "Hallo" and in my tactless way asked him if he knew anything about ecology. It was not a silly question, as long-term visitors tended to fall into one of two categories: those who thought they already knew a great deal, and those, who like myself, thought they knew next to nothing. Fager looked nettled, and told me sharply, "I don't know anything about ecology, but I *do* know how to do *research.*" Later I learned that he was a brilliant biochemist and mathematician. He had a Yale Ph.D., and while working in Chicago he had revived an earlier interest in biology and taken courses with Thomas Park. This had led to the award of a Merck senior Postdoctoral Fellowship, designed to help research workers move from one field into a different one. So Tom Park, who had sent over John Brereton, was again responsible for sending an interesting man to Elton, who judged Fager to be "next to George [P. H. Leslie], the topmost brain ever at the BAP."

Elton persuaded Fager to undertake the difficult study of a complex animal community. He felt it was necessary for such jobs to be tackled, but he seldom encountered anyone he thought capable of doing them.

Fager quickly showed that he did indeed know how to do research. He began by sampling the invertebrates living in small pieces of decaying wood, and then devised a method for testing some of the factors responsible for their presence. Decaying wood is wonderful stuff. It provides a host of animals with food, moisture, and cover. Fungi and other microorganisms break down cellulose and lignin into simpler substances that many animals can eat. And they also manufacture protein precursors and other subtle needs of some species. Then some of the animals meet their leasehold payments by opening up the wood physically by their boring and moving about, and thus make more of it available for fungi and a resurgence of the process. The difficulty is to make sense or science out of the apparently haphazard occurrences of the many species which turn up in different pieces of wood. Only by completely extracting the inhabitants of a number of comparable bits of the same wood; by accurately identifying them; and then by applying valid statistical treatment to the data, could anyone hope to elucidate what was going on. The Bureau was one of the few places where extraction apparatus for small animals was being developed. Wytham was one of the few pieces of woodland whose invertebrate inhabitants were recorded, and many of the larger forms were identified and preserved in the Boss's survey material. But Fager still faced formidable taxonomic problems which would frighten off most people sufficiently capable to undertake the work.

Small broken boughs picked up in the woodland, even if they are all from the same kind of tree and are lying in similar places, are at different stages of colonization because they fell at various times. In order to learn something about colonization, Fager needed a series of logs of identical age. But he could not wait for them to rot, so he invented artificial logs. These were boxes of oak packed with oak sawdust and "inoculated" with some decaying oak to get the fungi off to a good start. Holes were bored in the "bark" of the boxes to let invaders in. Elton described Fager's early results in his *The Pattern of Animal Communities* (pp. 297–98). The results and conclusions became Fager's contribution to the special number of the *Journal of Animal Ecology* issued to honor Elton on his retirement (63).

Tom Huxley worked on "The food of woodlice with reference to the break-down of dead plants" in a stony field near the Dell. He did not round off a sufficient parcel of results for a thesis, but his painstaking analysis of the grasses and other plants eaten by the woodlice in that part of Wytham was written up for the Nature Conservancy, and commented upon in PAC (p. 108) (51).

Bill Fager checking modified Tullgren apparatus for the extraction of inverte-brates. 1955. Photograph by Denys Kempson. Courtesy of Charles Elton.

In 1954 the Nature Conservancy selected an Oxford graduate, V. P. M. ("Pat") Lowe to become their rabbit expert, and it was logical that he should work under Southern at the Bureau. But the disease *Myxomatosis* began to wipe out Britain's rabbits before he could get to know much about their populations; Wytham's rabbits virtually disappeared. Pat was diverted to trapping experiments with small mammals, integrated with Southern's long-term study of tawny owls and their prey (134). He worked at the Bureau until 1957, when he joined the staff of the Con-servancy and carried out a fine study of the Red deer on the Isle of Rhum.

When Southern worked on the house mouse, one of the difficult en-vironments, from the control aspect, was the British corn-rick, that ar-chaic mouse paradise built by farmers to store grain through the winter. As there was little predation on rick mice, except from the occasional weasel, and from rats, the mouse inhabitants of a rick provided a popu-lation which could not be obtained by trapping. By placing a mouse-proof fence around a rick at the time of threshing, almost every mouse, of whatever age, could be captured. The reproduction of rick mice had been investigated during the war but all of the samples had been lumped together. It would obviously be interesting to examine the differences in age-structure and reproductive performance in separate rick populations. This opportunity was taken up by a Fulbright scholar from the University of Wisconsin, where studies of house mice had been fostered by one of the American wartime rodent control investigators, John T. Emlen, Jr. His student, Charles H. ("Chuck") Southwick, brought to the problem, not only the appropriate training from his Ph.D. work on the same animal but also the personal traits needed for dealing with the people he would depend upon for help—British farmers and laborers. Chuck was a quiet American, modest, courteous, and patient: As well as obtaining the house mice he needed, Chuck struck oil in finding the harvest mouse, *Micromys minutus* in a number of ricks (135).

Mick Southern was surprised and gratified by this as harvest mice had not been recorded in Oxfordshire or Berkshire for decades, and were thought to have vanished from the south of England. Their disappearance was attributed to changes in harvesting. In earlier times, when grain crops were manually scythed, the reapers would see and avoid cutting down the nests woven high in the stalks, leaving little patches standing in the fields. Mechanical harvesters, even when powered by horses, made no such concessions to Mother Nature. In the twentieth century, even the hedgerow refuges of such creatures were being destroyed as adjacent fields were merged for the more effective use of machines. Now, in 1954, Southwick began to find harvest mice in corn ricks, first a few at a time, and later (perhaps because the ricks had been standing for a longer time) more abundantly (136). Through Chuck, I acquired my first pet harvest mice, and we have remained joint legatees of Southern's love of small mammals.

Davies, Southwick, and Fager were replaced in 1955 by three very different but equally interesting people from overseas. As a frequent visitor to consult Southern about mouse matters, and for any other excuse I could invent to return to Oxford from London, I met and usually got to know the new generations of students and long-term visitors. One of the new recruits, a "loner" from Southern Rhodesia, managed to escape my recollection, although we must have met at tea. That was David Eccles, a graduate of Cape Town University. Elton described him as a brilliant taxonomist. He carried out a very full study of the fauna of a stream in Marley Wood Plantation, using emergence traps and other ingenious gear developed by D. K. The material was not worked up for publication, but that formidable task was later undertaken by C. A. Elbourn

David Eccles collecting emerging aquatic insects from a Wytham stream. 1956.
Photograph by Denys Kempson. Courtesy of Charles Elton.

(37). The other two new students were from the other side of the world:
Jiro Kikkawa from Japan and Kitty Paviour-Smith from New Zealand.
Jiro came to Oxford under the sponsorship of the British Council, which
in the previous years had sent his academic record to Elton. The printed
form had about forty boxes, all scored *Excellent,* except one. When the
Boss showed this impressive record to Fager, Bill protested, "But
Charles, you can't possibly accept a man who is only *Good* at Maritime
Law!" Jiro worked on the apparent trap-shyness of rodents, tackling the
appallingly complex problem of relating live-captures to the actual lives
of the animals: their movements, activity patterns, and their distribution. The trappability of individuals was found to be affected by their
social hierarchy, as well as by the spacing of traps and the frequency of
trapping (81). After a holiday visit to Fair Isle, Scotland, Jiro published
a note on the habitats of the field mice (80). Later in his career, he ac-
complished the remarkable feat of becoming a popular professor of zoology in an Australian university, the first native-born Japanese to break
the racial barrier erected by World War II in Australian academia.
Kitty Paviour-Smith was one of those students who won a place in

the Bureau through ability to overcome obstacles, as well as for the promise shown in their work. After graduating from Otago University, Kitty had studied the ecology of a salt marsh, and had sent her results to Hardy and Elton. They liked what they read, and in November 1954 Elton wrote to say that she could work in the Bureau if she could get the necessary funds for travel and subsistence. She obtained grants from a number of sources, including the New Zealand Federation of University Women, and the University Women's International Fellowship, and landed in Britain, after the long sea journey in September 1955. Meanwhile, Elton and Fager had thought about suitable projects and decided that the study of the fauna of tree bracket fungi was both worthwhile and feasible. They suggested the common small form *Polystictus versicolor,* but that winter proved to be so dry that the fungus became scarce. Kitty turned to a study of the inhabitants of the birch bracket fungus *Polyporus betulinus,* especially one of the beetles, *Cis bilamellatus* (11). She carried out an outstanding piece of detective work on this Australian species of beetle, which was first recorded in Britain in 1884, and first recorded in Wytham in 1935 (119). The collecting of beetles was very thorough in the early decades of this century, especially in the vicinity of Oxford, so this beetle probably arrived in Wytham very little before 1935. Kitty conducted a survey of its spread in Britain, and found that it had reached "the south and east coasts, south Wales and the Midlands up into Cheshire and Yorkshire." The first invaders probably arrived as stowaways in a sample of bracket fungus sent to Kew Herbarium for identification (118).

Kitty Paviour-Smith (later Mrs. H. N. Southern) remained at the Bureau until its demise in 1967, adding superlative data to the Wytham Biological Survey. When her D.Phil. work was completed, Elton had offered her the research officer post of his staff, a job he had been holding vacant since Amyan Macfadyen left to go into university teaching in 1956.

The Bureau had a full house in 1956, but with the departures of David Jenkins, Jiro Kikkawa, and Pat Lowe in 1957, there were spaces for a new selection. These were Janet Dawson, an Oxford graduate who had already done some part-time work for Chitty, Robin Newson, a recent graduate from "the other place," and Ian Efford, a graduate of Leicester University. With a DSIR grant and Chitty as her supervisor, Janet undertook to examine changes in the "quality" of voles.

This work was motivated not only by Chitty's unremitting efforts to explain vole "crashes" but also by the meteoric rise to prominence of a

Kitty Paviour-Smith collecting birch bracket-fungus. Wytham 1957. Photograph by Denys Kempson. Courtesy of Charles Elton.

newcomer to the field, John J. Christian. I was in the Bureau library when the issue of the *Journal of Mammalogy* containing Christian's first paper on the subject arrived.

Chitty was temporarily devastated by its confident application of Selye's ideas about the physiological disorders caused by stress to the phenomenon of small-mammal cycles; he felt beaten to the post, rather as Scott must have felt when he saw Amundsen's flag at the South Pole. Others must have pointed out, as I emphatically did, that the paper was an affirmation of a notion and, however brilliant and seductive the notion might be, it was not supported by evidence (18). But Dennis had been thinking along similar lines, and had not written anything about it because of the greater inhibitions he had about publishing unsubstantiated statements. With the passage of years, Christian searched hard for experimental evidence to back up the theory, and the more he learned about voles and contributed to the international discussion about their population dynamics, the more he moved away from his original position. Thirty-five years later, Chitty's hypothesis was more prominent in the ongoing debate.

BAP group on Wytham hill, 1957. From left: Janet Dawson, Mick Southern, Monte Lloyd, Ian Efford, and Denys Kempson. Photograph courtesy of Charles Elton.

Janet Dawson examined samples of voles from Wytham populations at different stages of their fluctuations in numbers, and other voles were maintained in the vole room as controls not subject to the same changes in their physical and social environment. Her experimental work was frustrated by the reluctance of the vole spleen to function as a calibrated barometer of population pressure. Stress comes, it seems, in many forms. A vole's spleen enlarged, for example, when it had fleas (15). Could one seriously attempt to weigh flea bites against vole bites? Voles brought into the lab and kept there under constant conditions showed similar changes in hemoglobin levels and reticulocytes to those still living in the field when breeding in the spring. But while growth was poor in Wytham populations after the crash, voles which should have been of the same "poor quality" did well in captivity. There were seasonal changes in the things measured, but no correlation, apparently, with changes in population density (29, 112).

Some of the voles used by Janet were from populations being studied by Robin Newson, with a DSIR grant. His thesis title, "The ecology of vole and mouse populations in different habitats," had a fortunate flexibility, as the small rodents in Wytham practically disappeared in 1957. The low levels of voles and mice at the outset of his study made trapping unrewarding, but it provided an opportunity to record their recovery (113). One interesting observation was that these small mammals, which do not normally breed in winter, did keep breeding throughout the winter of 1958–59, accelerating their return to "normal" levels of abundance. In 1958, not a single tawny owl produced young in Wytham, giving corroboration by a most efficient group of mouse hunters that the mice really were not there to be trapped. Robin mapped the vegetation meticulously in his study areas and found correlations between the nature of the cover and the occurrence of bank voles, but not with that of the wood mouse, a more rapid-moving and wider-ranging species. The odd little colonies of *Microtus* encountered here and there in Wytham woodland, seemed to survive there without any recruitment from the larger populations of the *Brachypodium* areas. There was a happy outcome to the cooperation between the two graduates; they married before their theses were completed.

The third student starting work on a D.Phil. in 1957, Ian Efford, had attended the BAP field ecology course in the previous year, and had impressed Elton with his grasp of mite taxonomy. During the tenure of his two-year grant from the Nature Conservancy he showed such ability in that difficult field that Elton arranged for him to stay on for a third year to help with the classification of the Wytham collections.

On arrival, Ian was pointed in the direction of *Microtus* work under Dennis Chitty, as the Boss never liked to supervise more than one student at a time, and he was already advising Kitty Paviour-Smith. But Ian was not very keen on that idea, and after some prolonged debate, he prevailed in his wish to work on water mites. Dennis was not reconciled to the task of supervising research in a field in which he had no experience, but he conscientiously committed to memory the outlandish names of the water mites occurring in the pond where Ian started to work.

After a time, the focus of the study shifted to the fauna of the springs studied by Eccles, who was still at the Bureau and gave valuable advice, both with taxonomy and techniques. Thus, when Dennis received a progress report, he was alarmed to find that Efford appeared to have got the names and combinations all wrong.

The main work was on the tricky life-history of a relict cold-water mite, *Feltria romijni* (35). Its larvae parasitize a minute Chironomid fly,

Ken Marsland's cartoon commemorating Ian Efford's nightmare, 1958. Courtesy of Ken Marsland.

which was caught in the emergence traps Eccles had developed. The mite proved to be far more numerous than any previously studied population, with its greatest density in October 1958 of almost 1,400 per square meter (a density similar to that of terrestrial mites) (36). The heavy mortality which prevents any species from overrunning the earth appears to occur when the larvae are on the host fly, and among the young adults during the winter. Efford also studied the host fly in some detail (34). In his spare time (he was already a well-organized man), he rediscovered in Wytham Park an ancient Crustacean, *Bathynella*, previously known from subterranean gravels and caves in southern England. This form is related to the living fossil *Anaspides tasmaniae*, which lives in the mountain streams and tarns of Tasmania (33).

In the same year, Thomas Park sent from Chicago another of his bright young men, Monte Lloyd, who had a National Science Foundation postdoctoral grant for two years. Monte worked on the distribution of invertebrates, especially centipedes, in woodland litter (98). After the first year, he had impressed Elton so much with his fresh ideas that a Nature Conservancy grant was obtained to enable him to stay on. It is said of

Monte that he was capable of such immersion in his work that he once went home for a meal twenty-four hours late, and was not aware he had "lost" a day. He worked on a new concept, *mean crowding,* an attempt to assess the degree of crowding experienced by a typical member of the population, making allowance for the patchiness of the distribution of litter dwellers (99). Such a concept attracted George Leslie's attention too, and soon Monte and George were working on a method of estimating the standard error of mean crowding. The chief statistician of the Nature Conservancy, J. G. Skellam, also became involved in these esoteric matters.

Two other Americans visitors should be mentioned, as they visited for substantial periods. Frank Pitelka, who had done a vast amount of work on lemmings came from the Museum of Vertebrate Zoology in Berkeley for a sabbatical year. He carried out some field work in Wytham but did not publish the results. The other was Eugene Odum, author of the very successful textbook on ecology. He did not get off to a very good start with the Boss. He breezed into the museum and dropped the manuscript of his second edition on Elton's desk, rather expecting it to be eagerly perused by the senior staff. It stayed there undisturbed for some time, as the Bureau people were very busy with their own interests. But then Odum proceeded to win everyone's regard through his enthusiasm and energy. In only four months, he marshalled the resources he needed from four university departments and carried out a study of yellow ants in collaboration with A. J. Pontin of the Hope Department of Entomology. Using marking-recapture methods and radioactive tracers, they measured the population (about 40,000) of a nest on the Bowling Alley in Wytham (116).

As we have seen, an interest in lemming population cycles was one of the factors leading to the creation of the Bureau, but after the publication of *Voles, Mice and Lemmings,* Elton's personal research interests lay elsewhere. It gave him great vicarious pleasure, nevertheless, to arrange for Robin Newson and his wife, Janet, to visit South Norway in July 1959, to observe the predicted superabundance of lemmings. The Newsons were accompanied by Dr. H. J. T. Gall of the University of Chicago. A year later, the Newsons went back to observe the predicted crash.

In 1960, two new long-term visitors arrived, one from a village in Ayrshire, by way of Glasgow University, and the other from a bush town in South Australia, by way of Adelaide University. The Scot, W. W. Murdoch, with Elton's advice and supervision, undertook a study of water-margin-living Carabid beetles. The main subject was *Agonium fuligi-*

Bill Murdoch collecting beetles in the marshy verge of the lake in Blenheim
Park at Woodstock. 1961. Photograph by Denys Kempson. Courtesy of Charles
Elton.

nosum, abundant in marshes and streams, now studied in Wytham and
in the lovely grounds of Blenheim Palace. The population dynamics
were investigated through regular samplings by means of pit-fall traps,
hand collecting, and extraction from marsh litter and sallow logs. At the
same time other Carabids were collected, and the life histories of some
"wet" and "dry" forms were worked out (109, 110).

The Australian student, M. E. B. "Mike" Smyth, was a Rhodes scholar
from Adelaide, where the zoology department, under the chairmanship
of H. G. "Andy" Andrewartha, taught with heavy emphasis on methodology and statistics. Mike was initially attracted to the Bureau by the
prospect of working with Chitty, whose hypothesis about the mecha-
nism of vole cycles intrigued him. But by the time he got to Oxford,
Dennis was packing to go to British Columbia, so Mike was obliged to
work with Mick Southern. He set out, nevertheless, to undertake a project suitable for one of Chitty's students: "The effect of varying abundance
on the population dynamics of rodents, with special reference to the
Bank vole, *Clethrionomys glareolus.*" His ideas to be tested were (1) if sufficient animals are removed from a population approaching its peak

density, its crash will be averted, and (2) if animals are removed from a declining population, the decline "should not be arrested until it has run its course." He trapped and removed the bank voles and wood mice from an area in Marley Wood, Wytham. Other areas were monitored as controls.

There were no definitive results. Survival rates were higher in the area from which animals were removed, but that could have been due to unmeasured differences in the habitat, especially in food supply (129). With the resources at his disposal, detailed analyses of such factors could not be made. He concluded: "In view of the inadequacy of the experiment I cannot claim to have tested Chitty's hypothesis. But although I have not properly tested it, I now doubt it. I doubt it mainly because I am not yet convinced by the most important point in his argument, that all declines (or most, calamities aside) belong to a single class of events. . . . Until Chitty gives better evidence for his claim, his hypothesis is needlessly complex." Mike's remarks were in the style of the Adelaide school of the time; tending to be needlessly polemic, and by overstating the views of others, making them easier to knock over.

Another aspect of Smyth's investigation related population changes to cycles in the supply of food from trees. It seemed that a good acorn crop encouraged extension of the normal breeding season into winter, so he was obliged to refer back to the views of Baker and Ranson, that the breeding season represents, "an adaptive response of stopping and starting reproduction when the temperature and food alter with the season beyond certain limits" (128).

In spite of Mike's obvious disappointment with his time at the Bureau, there was a happy and positive outcome. When Marjorie Nichols, the staunchly loyal librarian had left to marry and raise a family, she had been replaced by an attractive young lady named Janet Halliday. Mike took back to Adelaide not only an Oxford D.Phil. but also Janet as his wife, and Elton was again hunting for a librarian. Mike's fruitful career and happy family life were both cut short by his premature death from cancer.

Dennis Chitty left the Bureau in September 1961 because he had decided he wanted to teach. It was a gamble, he felt, because he did not know whether or not he would be any good at it. But he felt he had been in research for twenty-six years without interruption, and could now do what so many colleagues felt unable to do, devote himself full-time to his students, "who God knows, get a fairly raw deal in large universities staffed by people whose careers depend upon publishing original re-

Informal lunch at the BAP, 1962. From left, Bill Murdoch, Tony Dunsford (staff), Francis Evans, Mike Smyth (reading tea-leaves), Janet Halliday, and Cliff Elbourn. Courtesy of Charles Elton.

search." It was appropriate that the Chittys returned to Canada, where Dennis became a professor in the University of British Columbia. He was true to his resolve to give all of his time to his students, devoting his summers to graduate students, and only getting back to personal research in 1972 "as a low priority occupation." Helen Chitty had collaborated with Dennis for most of his career, taking some years off to devote to raising their children, and returning to research on a part-time basis. So the departure of the Chittys reduced the emphasis on vertebrate studies in the Bureau. Although the *Microtus* research came to a pause, trapping studies of small mammals continued under Mick Southern's supervision. From now on, a greater proportion of the Bureau's modest resources would be devoted to invertebrates.

The Boss was largely occupied with the completion of his scholarly book on the patterns of animal communities, and with gathering specimens and data for the Wytham Biological Survey. For six months of 1960, he had acquired a research assistant, J. W. Hurry, a Cambridge graduate, to help with the survey. But he found Hurry needed too much supervision in meeting his exacting standards, especially as Hurry some-

Three co-authors admiring the proofs of their paper on a new extraction apparatus for leaf-litter fauna. 1962. From left, Ray Ghelardi, Monte Lloyd, and Denys Kempson. Courtesy of Charles Elton.

times spelled Latin names backwards. Hurry subsequently became a
very successful teacher in the British Open University.

The interesting work begun by Monte Lloyd on sampling litter invertebrates was still under way in 1961. Dr. Ray Ghelardi had been sent to
the Bureau by Bill Fager, who was then at the Scripps Institute of Ocean-
ography. Ray teamed up with Monte and D. K. to develop a new kind of
high-efficiency multiple-unit extractor for small samples of litter. One of
the problems which had frustrated Monte's earlier attempts to measure
populations was to get enough samples without the pattern being con-
fused by the passage of time. With the new equipment, it was possible
to extract the animals from a number of samples simultaneously (79).
Monte took all of the material back to Chicago to work up, got immersed in other things, and it was never published. When the work had
to stop, the team had begun looking at hexagonal sampling areas, a brilliant breakaway from the use of quadrats, a cultural inheritance from
botanists.

The loss of the Chittys was bound to result in a change in the direction

Professor Peter Larkin on fieldwork in Wytham, 1961. Photograph by Denys Kempson. Courtesy of Charles Elton.

of research, especially that of new senior students, and inevitably there would be less interest in *Microtus*, which had become an intellectual quicksand. But the mainstream of ideas still flowed quietly from the Boss, and losses can be made good by the right appointments. Kitty Paviour-Smith became a staff member and took on the supervision of some D.Phil. students. Peter Larkin, now a professor at the University of British Columbia, came for a sabbatical year and adapted Fager's artificial log technique for the study of faunas colonizing rotting oak boughs at different heights above the ground. This project was carried out with C. A. B. Elbourn, who worked as Elton's assistant on the Wytham Biological Survey from 1961 until Elton's retirement in 1967 (85).

The loss which could not be made good was that of the Bureau's best friend, Professor Sir Alister Hardy. Hardy had retired early from the Linacre Chair to allow his successor to get in place before the planning of the proposed new quarters for the department was too far advanced for him to have any input. Although Hardy stayed on as head of the Department of Zoological Field Studies, that was not an influential post, having been invented as an administrative folder in which to place the BAP and the Edward Grey Institute. J. W. S. Pringle, a molecular biologist, as the new Linacre professor, promptly pronounced that he did not believe in having small, independent research institutes within a department of the university. He felt he had been appointed to "clean up this mess."

6   In Wytham Woods

On the fourth day of March 1949, King George the Sixth, "by the Grace of God, of Great Britain, Ireland and the British Dominions beyond the Seas, King, Defender of the Faith, by virtue of His Royal Prerogative and of all other powers enabling Him on that behalf, did of His special grace, certain knowledge and mere motion," and so on and so on, "grant, will, ordain, constitute and declare" that fifteen distinguished gentlemen, including Charles Elton, Esq., M.A., should be the founding fathers of a body to be known as the Nature Conservancy. The functions of this body were (1) to provide scientific advice on the conservation and control of the natural flora and fauna of Great Britain; (2) to establish and maintain and manage nature reserves in Great Britain, including the maintenance of physical features of scientific interest; and (3) to organize and develop related research and scientific services.

The first director of the conservancy was "Captain" Cyril Diver. When Diver retired, a prominent civil servant and amateur ornithologist, E. M. Nicholson, was moved into the position from the Treasury. It is in the Treasury that British power dwells, and it was said at the time that the Treasury "Old Boys" had thought that by giving Max Nicholson the Nature Conservancy, they were putting him where he could not build an empire. They were mistaken. Under his dynamic leadership and articulate advocacy, the Nature Conservancy grew rapidly in size and influence (and budget).

Elton, Diver, and Nicholson were all members of the founding Wild Life Conservation Committee (the "Huxley Committee") set up by the Ministry of Reconstruction. The great botanist A. G. Tansley was vice-chairman of that committee, and when Julian Huxley went off to head UNESCO in 1946 Tansley became chairman, and hence chairman of the new body. Before the war, Elton was a member of the Committee of British Ecologists. This committee largely worked on setting up "ecological reserves" within Forestry Commission properties, on monitoring changes in plant and animal life therein, and on some experimental studies. During the war, the British Ecological Society set up an enquiry (1943) to investigate the need for British nature reserves on a broader scale; to preserve pieces of habitats of special interest, wherever they might be in Britain. Tansley had been chairman of that committee, and Elton had represented the animal side. Another influence leading to the formation of the conservancy was the Nature Reserves Investigation Committee (1942). Some members of that were also members of Tansley's committee. The conservancy was even more inbred: at first, it had functions which related it to the Agricultural Research Council. These biopolitical matters are not relevant to the present work, and they are recorded in detail by John Sheail, but they show that Elton had a long-standing interest in conservation. The activities of these various committees were eclipsed by the Huxley Report. While that report was being digested, the members of the Huxley committee were invited by the Minister of Town and Country Planning to act as "Wild Life Scientific Advisors." For a year, this group constituted the embryonic Nature Conservancy. "Elton, for example, made a two-day visit to the proposed reserve at Ainsdale in Lancashire, where he made a vegetation map on the basis of R.A.F. photographs and a ground survey. This made it possible to determine the precise boundaries of the protected reserve for the first time" (123).

Elton was a reluctant committee man, and he became more reluctant as time passed. He hated to "dress up like a silly ass" and waste a precious day sitting around a table in London. His secretary, Marie Gibbs, told me that he used to produce doodles of recumbent rabbits holding their paws over their eyes in despair. But he liked to inject his ideas into conservation matters, and while Cyril Diver was running the conservancy, used to turn up at meetings. It is safe to assume that he exerted much more influence on policy than he liked to admit. When we were comparing notes on our conduct at committee meetings, and I confessed that whenever I felt I was about to lose my temper, I excused myself and went to the men's room to cool off, he said that he only recalled "getting hot" and making an impassioned speech on one occasion. "Tansley was in the chair," he said, "and this report came before us about spraying roadside vegetation with herbicide. The firm which wanted to do business with 14 Local Authorities had put out a brochure, and one statement concerned getting rid of the English Rose, which was referred to as a *noxious weed*. 'Do you realize, gentlemen,' I said, 'that you are about to allow a contractor to destroy the basis of much of our English poetry—the wildflowers of our country, many of which are now to be found *only* in the hedgerows and beside the roads?' Tansley sat up and said, 'H'm.' Then he turned to Diver and said, 'Can't you do something about this, Diver?'"

The problem was resolved through the establishment network. Cyril

Diver had a quiet word with Dr. China, Keeper of Entomology at the British Museum (Natural History), and the top bug man in the country. China's main asset in this matter was that he happened to be the brother of a senior executive in the company which had made the spraying proposal. Many awkward situations have been thus resolved during luncheon at the Athenaeum. The proposal was dropped.

Although the Nature Conservancy grew and flowered under Max Nicholson's administration, it was dismembered after he retired and its various divisions were sent off in various directions. The research staff came under the control of a Natural Environment Research Council (NERC) and were placed in the Institute for Terrestrial Ecology. So it came to pass that Nicholson, who had given up Elton's cause for lost in 1964, also lived to see the organization he had created put to death by lesser men.

Elton's influence on nature-conservation policy was evident in the high priority given to biological surveys of the lands under the convervancy's control. He urged that stock-taking be regarded as a normal duty of reserve wardens, and that they record, through the keeping of diaries, dynamic matters as well as mere lists. But he did not expect people with such heavy and diverse responsibilities to undertake sustained or in-depth studies of communities in their care. J. Sheail wrote:

> According to Mr. Elton, a member of the (Huxley) Committee, ecological surveys were at the core of all management programmes. They had to be undertaken at three levels; by the various university departments, the scientists in charge of the reserves, and by the wardens of the reserves. While the warden, "should be firmly protected against too much paper work," he had a duty to keep weather records, phenological records of selected species, and a daily log book of other observations (123).

The development of Elton's ideas about ecological surveys is set out in the introduction to his paper with Richard Miller, on the classification of habitats (56). His exposition on the subject in *The Pattern of Animal Communities* is more complete, but the wide-ranging and sometimes whimsical allusions are liable to sidetrack anyone less scholarly than himself. His ideas about community ecology, the ecological grouping of species, that is, the pattern of organic nature, owed much to those of Victor Shelford (124). Then there were his own studies, as a student, on the inhabitants of marl pits and sand dunes in the Liverpool area, where he was struck by the upper and lower limits reached by some species (40). "Working in the Arctic; the exhaustive examination of plant and animal communities on Spitsbergen, *made one see ecological processes in full swing.*" Although these observations were not quantitative, they gave him insights into the complexities of the relationships between species, even in an austere climate. He had carried out, with O. W. Richards, studies on the stabilization and colonization of mud in the Oxford canal. The work was not published, but it "left him with a vivid impression of the power of vegetative, and resulting microclimatic, conditions to alter the composition of animal communities."

The theories of Lotka, Volterra, Thompson, Nicholson, and others were really saying that *the limitation of numbers must be brought about primarily by biotic relationships.* Of the five kinds of relationships possible, four of them, leaving aside intraspecific competition, concerned relationships between different species. It followed, therefore, that studying species associations was necessary for understanding the population dynamics of any one species. This was an important philosophical change from botanical ecology, as the properties, the "elasticity," as it were, of the biological network itself became paramount rather than factors like climate (47).

Elton's interest in surveys continued during the war, but because of the pressure of the tasks at hand it could only be a planning exercise. He had intended to undertake a regional survey, taking Oxfordshire, Berkshire, and Buckinghamshire, and covering all kinds of habitats and their occupants. This had to be severely pruned for the sake of practicality, and he settled for doing an intensive job on Wytham. After some years' work, limiting the survey to Wytham Estate did not seem so restrictive after all, for that small piece of Britain proved to have living within it a large proportion of the fauna of the country.

The first serious reconnaissance of Wytham was carried out in April 1943. The development of the Wytham Biological Survey (WBS), its philosophy, methodology, and ecological objectives went on for another ten years, the most productive being probably 1952, when Richard Miller and Brenda McPherson provided the Boss with an amusing sounding-board and congenial company. Although the form of the survey was crystallized in that first decade, and the habitat classification system published, the painstaking gathering of specimens and facts about them continued. The many visitors from overseas were taken for a walk in Wytham rather than permitted to waste precious time sitting and chatting politely in the Bureau. This showed them something of the English countryside in the company of a superb naturalist, and provided healthy exercise. But most important, it produced more records and specimens for the WBS.

Almost every thesis arising out of work in Wytham begins with a description of "the study area." For some students, an account of a small area of the estate has been sufficient, but for others who worked in a number of different habitats there, it has seemed relevant to write elegant descriptions in the Oxford manner. If they were not Oxbridge types, who had learned to write good prose through the college tutorial system, they learned belatedly from the Boss or Mick or Dennis—whoever had the job of pulling their thesis into better shape, and editing out the unnecessary jargon. The Boss quoted his father only once, and I have tried always to apply the rule; *it is almost impossible not to improve a piece of English prose by taking something out!* As the product of a technical school, and of a small university which followed the Scottish tradition of giving the students so much obligatory work to do that they had no time to get into mischief, I had no idea how to write succinctly, so my thesis was vastly improved by Southern's advice. He was a classics graduate before he decided to read zoology, and while doing his second Oxford degree he wrote a column for the local newspaper (in Berkshire dialect!). When I tried to describe Wytham in my own words for this chapter, I failed badly. In trying to describe what it felt like to hear a cuckoo, to see the lacy foliage of beeches in spring, all of my attempts seemed mawkish as well as irrelevant. But the history of the Bureau of Animal Population became so intertwined with the advancement of knowledge of Wytham Estate, that it is necessary to impart a feeling for the spirit of the place. Elton has said that it is necessary to do fieldwork in order to have time to *think;* it has always given me time to *feel.* So I tried to find the best description, and asked Elton's opinion. He chose that by Valerie Todd, who studied harvestmen (Opiliones), and was based in the Hope Department of Entomology (140). Miss Todd certainly gave an excellent factual description, but not the kind of evocation I wanted.

I thought my purpose could best be served by borrowing some of the Boss's phrases:

> Just before it flows through Oxford, the River Thames
> makes a great loop to the north, east and then south,
> running in a wide alluvial valley that still gets partly
> flooded in winter. In this shepherd's crook of a river lie

two hills closely connected by a col . . . rising about 300
feet above the river level. . . . Inside the river loop the
University now possesses 3400 acres of land—or a little
over five square miles—of which the wooded part of
the hills themselves with some enclaves of park and arable fields form about a third. . . . The northern one is
Wytham Hill, the southern one is Seacourt Hill; but I
shall refer to the whole block . . . as Wytham Hill, as
indeed is general practice nowadays except among ge-
ologists. The slopes and top of the hill have nearly a
thousand acres of scrub and woodland . . . seen from a
distance the hill lies low and is dark with woodland,
except for the Park (a long valley between the two parts
of the hill, with large scattered trees of English elm, oak
and ash), and the curious arable field and grassy enclaves near the top. . . . On entering the woods one
passes into an irregular mixture of oak and ash and elm
and sycamore, not always in dense or very high canopy. . . . On some parts of the hill there are stretches of
bracken or oases of bracken-scrub savannah. . . . Under
the canopy of mixed deciduous wood a great deal of the
ground is hidden by an extensive green sheet of dog's
mercury less than a foot high. . . . Under the beeches,
however, there is usually pure leaf litter. . . . Here and
there one encounters small fast-running streams in
heavy shade and quite devoid of green plants. . . . Or
one walks into the edge of a woodland marsh with vegetation almost head-high in summer. . . . On top of the
hill the open parts on shallower limestone soils are
graced with a particularly beautiful series of mead-
ows. . . . Down by the muddy River Thames there is a
special riparian zone, half meadow and half marsh. . . .
The Hill has many variations on these themes. . . . (51)

This beautiful piece of England, probably the most intensively studied
piece of land in the world, became the property of Oxford University in
1943. A wealthy South African, R. W. Schumacher, had decided to become an English country squire. He purchased a suitable estate with its
village of quaint thatched cottages and local public house (suitably
named "The White Hart"), and, to show that he did not do things by
halves, he changed his name to "ffennell." By taking an active part in
the Home Guard, he also acquired the honorary rank of Colonel. Colonel and Mrs. ffennell had only one child, a tall dark girl who roamed the

lovely estate like an exiled princess, adopted a band of gypsies who were permitted to camp on the estate, and suddenly, after a short illness, died.

When Colonel ffennell died in 1943, his wife gave Wytham Woods to Oxford University, with two stipulations, (1) that "every care be taken to preserve the woods in their present state of natural beauty" and (2) that the university "use them for the instruction of suitable students and provide facilities for research." At the time of making the gift, Mrs. ffennell gave the university the opportunity to buy the other 2,000 acres of the estate, including the Abbey and the village. She retained the right to live in the Abbey for the remainder of her life.

In the early years of management by the university, there were divergent points of view, as a number of departments were involved. From the ecologists' point of view, the most dangerous managers were the foresters. They liked neat and tidy stands of trees. Fallen timber, even fallen twigs were, in their eyes, havens for pests and parasites. For Elton especially, this "housekeeping" removed vital minor habitats for a multitude of species, and he became the "champion of the rotten logs." The more he came to appreciate the richness of the fauna of rotting wood, the firmer the stand he took, and after a time, his view prevailed. Professor Alister Hardy reported, in his last annual review of the work of the Department of Zoological Field Studies (1963):

> The new modified and mild programme of woodland management in Wytham Woods decided recently by the University will ensure the continued survival of a variety of habitats with their fauna. This is of the utmost value for continued research on ecology in the future. The Bureau is much indebted to the University Land Agent, Major J. R. Mills, for his active cooperation with research workers there.

The interests of the various departments waxed and waned, and management was largely passive, leaving nature to take its course. Without active intervention, many habitats inevitably changed drastically. When the rabbit practically disappeared, much of Rough Common, a vole habitat dominated by *Brachypodium*, became colonized first by lush grasses, and then by scrub and trees. As living landmarks vanished, some of the traditional names became inexplicable; when I first walked in Wytham, the "Five Sisters" who were the silent sentinels at one end of "The Singing Way" were no longer five in number. In 1980, only two were still standing, and they would soon join their sisters as rotting logs. The young conifer plantations in which I first captured a pygmy shrew were soon sterile stands of timber, not only wiping out the lush small-mammal habitat, but also blocking some of the long misty prospects. Perhaps the most constant and unchanging phenomenon during four decades and more was a gentleman in a cloth cap, carrying a collecting net and a haversack, walking carefully through the woods. Seven years after Elton's retirement, the management of the estate was taken out of the capable hands of the Estate Management Department, and made the responsibility of a committee of "fat cats" who never went into it and had no ideas about it. Elton encouraged some of the younger people to get involved. Just as voles undergo fluctuations in numbers, so do the activities of man, and soon the estate was in capable hands again.

If Mr. Schumacher had not become enamored of the English way of life and undergone the transformation into Colonel ffennell, the studies of the staff and visitors in the Bureau, and in a number of other departments of the university, would have differed in detail, but the main threads of ideas and research would have been much the same. Elton would have studied and analyzed the inhabitants of a different piece of England; Lack would have studied titmice in another place perhaps, and Varley might have counted oak moth caterpillars somewhere else. The reviews of Elton's last book tended to concentrate on the importance of Wytham. But the ideas cloaked in assemblages of examples and images of nature will outlast Wytham Woods, for they have changed our way of looking at the world. The main ideas expressed in the book were so immersed in the body of facts and references that some reviewers missed them altogether. I thought it a pity that the main ideas set out in the last chapter were not anticipated in the first. Then, instead of serving to remind the few intellectual alpinists where they have climbed, they would have given the many who gave up in the foothills a map to help them with the more difficult parts of the ascent. The difference between the basic biological survey, now a familiar concept to anyone interested in natural history, and Elton's ecological survey, which seeks to discover the dynamic processes within an animal community, is like the difference between elementary algebra and the differential calculus. Mapping the species assemblages and their biotic network, the natural history and statistical stage, although difficult and still at an elementary level, "is only a jumping off stage for the study of the community in action, considered as a system of events and processes operating in mixed populations."

> The early organic chemist must have felt very much as
> the animal ecologist often does now. It must have
> seemed very dangerous and vain to look beyond, say
> formaldehyde or a benzene ring, towards higher com-
> plexities. Yet organic chemists now seem to move with
> great confidence amidst compounds of complicated
> structure, built together by steady research upon chem-
> ical compounds and on the general properties of such
> systems. Perhaps it is fairer to assess the present posi-
> tion of the animal ecologist with that of the *inorganic*
> chemist at the time of Dalton!(51)

In 1927, Elton introduced the concept *pyramid of numbers* in bold type
at the beginning of a section in his *Animal Ecology.* In 1966, its corollary
idea is slipped into the text, half-way down page 189: "This phenomenon might be termed *'The Inverse Pyramid of Habitats,'* that is, the higher
the consumer level, the fewer and larger the animals, but also the habitat range. . . . The principle applies chiefly to food chains based upon
invertebrates, which form the great mass of animal life." (Lack would
have introduced this idea with the heading to a section, and Bodenheimer would have headed a new chapter, "Ein neues Prinzip.") But the
most important new metaphor is tucked away in the text on page 213:
"Askew's summary of the known British host-galls of the Chalcid parasites, after omission of the ten species not found at Wytham . . . shows
that 13 are specific to one gall host, 6 to 2−4 galls, 4 to 10−14 galls and
two others to 16−19 galls. It is to this sort of relationship between
centres of action within an ecosystem, or rather to 'populations' of
them, that I shall apply the term 'girder system,' a metaphor that contains both a fact and a hypothesis; the fact that there are strong biotic
links between different gall communities, and the hypothesis that such
links may give strength and stability to the whole interspersed population complex in galls." The *girder system* concept is stated as a generality
in the final chapter:

> It is reasonable to suggest, though it cannot yet be
> proved, that the interlocking of biotic connections right
> through a terrestrial ecosystem is one of the chief rea-
> sons for this (stabilizing) property. In turn, if this is
> really the case, the girder system must have some effect
> either in controlling numbers or at any rate damping
> down fluctuations or slowing down departures from
> the norm.(51)

We are comfortable with the idea of wheels within wheels, because we can picture in our minds such mechanical clockwork, but the living world has complexities of a higher order. Perhaps *webs within webs* is closer to a mental picture of the tracery of girders holding up a community structure. And the system has to be thought of as existing in four dimensions, because the threads of the webs are spasmodically being broken and reestablished with every ecological event. And just as the modern physicists have robbed us of our idea of solid matter, by telling us that it consists mainly of spaces between charged particles, so must the modern ecologist confront the idea that the complex biotic web also consists mainly of emptiness; much of the time the animals are not interacting, for living creatures rest a lot. The more complex the structure, the more elasticity it must have, and the less any one ecological event, or absence of an event, will change it overall.

For almost the whole time that Elton was engaged in his field work in Wytham, there was a long-term study under way by Mick Southern; a classic study of the tawny owl and its prey (131). The owl and mouse project was started in 1947 and wound down in 1959. From October 1948 until August 1952, there was a laborious capture-mark-recapture study of the wood mice and bank voles that were the tawny owl's main sources of food:

> During these years an area of 234 acres (94.7 ha), nearly a fifth of the whole estate, was gridded in 1-acre squares and two traps were placed in each square every second month. Positions in each square were chosen from a table of random numbers. The captured animals were marked individually with numbered monel metal leg rings, and over the four years, some 10,000 captures were made. (134)

Working as Southern's field assistant was the best possible way to learn about the English countryside and its plants and animals. Mick possessed extraordinary auditory acuity; he could pick up a faint twittering in a distant bush and tell you what kind of tiny bird would emerge. I especially remember him introducing me to some easy ones. My first question was about some sounds in Great Wood, in the early spring of 1949. They came from a chiff-chaff. And some months later, as we drove up from Marley Wood gate, he suddenly stopped the Land Rover and said, "Just listen to that!" From the hedge beside the road there came a rich contralto serenade of a nightingale. From time to time

I was able to pay him back with bits of knowledge missing from his considerable store. Finding monel-metal rings in tawny owl pellets could be simplified, I pointed out, by passing a powerful magnet over them. He had not realized that monel metal, being mainly nickel, was attracted to a magnet.

There was a great deal of physical work; carrying a metal ladder into all sorts of situations, climbing ancient trees to inspect possible nesting sites, making daily inspections of owl chicks, observing the owls by night from a hide, using a sniperscope (the very same one used for watching rats in Giles's piggery), as well as trapping and marking many, many small animals. In summer there was hot and sweaty work slashing the bracken that grew higher than our heads in the open glades, to clear the trap lines. After some hours of that, plagued by flies almost as bad as those in the Australian bush, it was pleasant to break for lunch and go to the cool haven of the Trout Inn at Godstow, where a large benign barman named Ernest ministered to the spiritual needs of a small congregation of local businessmen and yokels, students, and theatrical personalities. The unofficial president of this informal club was W. F. "Buzz" Burrows, the retired publisher of the student newspaper, "Isis." Buzz had two hobbies that filled his life; drinking at the Trout and acting as honorary game warden at Wytham. He could always be relied upon to greet us with a hearty, "Well! What are you taking forrit?" The people from the Bureau were his favorite guests, especially Mick, who lovingly coped with the enormous pints of Worthington that I was unable to tackle. He was obviously disgusted that I always ordered lemonade. It became embarrassing to accept so many free "rounds" without being permitted to pick up the tab for a change. One of the affable regulars, who used to cycle across the Port Meadow for a daily pint was the owner of an old-fashioned department store in St. Ebbes. His son, Bill Potter, became one of Mick's most valued summer helpers (132).

The community structure of the population utilizing the moist habitat at Godstow, on the bank of the River Isis, had much in common with communities in Wytham Woods. Some forms were always to be found there at certain times of the day; others were seasonal; and some were residents of quite different habitats who happened to be in the vicinity and were attracted by the nectars available there. With the passage of years, it changed, like all habitats. Its fame spread and too many became aware of its charms. It changed hands, prospered, grew, and became an enlarged imitation of its former self. When I last paid the Trout Inn a visit it was serving a smorgasbord lunch to busloads of tourists and was quite unsuitable as a habitat for its former visitors.

The publication of Elton's ideas about animal communities did not stimulate a surge of work of the kind that he believed to be important. Indeed, such work has been largely ignored by British ecologists, not only because they were already committed in other fields but because of the taxonomic and logistic problems inherent in such long-term studies. His work is better known, and probably better understood, in European circles. In the United States seekers after the underlying principles thought to be capable of discernment within the apparent complexity embraced the techniques and algebraic dialect of simpler sciences. When the Oxford Animal Ecology Research Group considered how they could comply with Elton's desire that they use and continue his survey in Wytham they came to the conclusion that the only way *was to clone Elton.* The data he had assembled so far were not amenable to computer treatment. But one of the outside statisticians consulted at that time, when asked about it twenty years later, offered the thought that the problem might be worth looking at again, bearing in mind the greatly increased facilities for analysis and the range of new techniques which might be appropriate for the material. Perhaps the last words of this chapter are most appropriate coming from a theologian. Peter Levi, S. J., reviewing *The Pattern of Animal Communities* in *The Tablet* commented:

> If any decently educated man towards the year 2060 is asked what was going on of importance in Oxford a hundred years ago he can properly start his answer by speaking about the fundamental research into untamed nature which was going on under Mr. Elton in Wytham.

7   The End of the Bureau

In February 1962, Charles Elton made a belated effort to raise the traditionally low profile of the Bureau of Animal Population within the university community. The staff presented posters and specimens to illustrate the main lines of research: small-mammal trapping, owl studies, marking methods, extraction apparatus, and material about the Wytham Biological Survey. Faculty members and others were invited to an illustrated lecture by Elton describing the origin and development of the Bureau. It has been quoted above. The conclusion was an expression of the Boss's desire that the Bureau continue to exist: "Professor Goodrich once remarked to me, at a rather bad time: 'you know, there's nothing so permanent as a temporary arrangement.' I hope the reverse will never be the case. After another thirty years, I may not be taking much interest in things, as I should, if I lived, then be 92! But even then I would like to know that the BAP was still going strong as an institute, and that its research had advanced so far that I could hardly understand anything that they were doing."

In the summer of 1962, Monte Lloyd and Ray Ghelardi went back to the United States to take up academic appointments, and Peter Larkin returned to Canada from his sabbatical leave. There was no intake of new graduate students that year, but there were two notable long-term visitors. Francis Evans, who had been the Bureau's first graduate student, and had become director of the University of Michigan's Laboratory of Vertebrate Biology, returned for a sabbatical year, and Harold Trapido, a senior staff biologist with the Rockefeller Foundation, came to spend more than a year, mainly working up his field data from investigations of tick-borne diseases in India. Trapido had a look at the parasites of Wytham's small mammals, and to Elton's surprise, for he had studied them intensively himself, discovered a new parasite of the wood mouse—a Trematode living in the nasal cavity.

Three new students began research in Wytham in 1963: C. P. Mathews, from Swarthmore College, Pennsylvania; S. L. Sutton, an Oxford graduate; and C. H. S. "Chris" Watts, from Adelaide University. Elton also recruited a new research officer, Dr. John Whittaker, whom he had

Professor Francis Evans, the first graduate student at the BAP, during a sabbatical visit in 1962. Photograph by Denys Kempson. Courtesy of Charles Elton.

met and formed a good opinion of while visiting the Nature Conservancy's Moor House Nature Reserve in the Pennines. Professor Cragg used to take his students there from Durham, and John was one of them, working on insects.

Mathews worked on the fauna of a small stream in Wytham, mainly studying the abundant crustacean *Gammarus pulex*, its population dynamics, migration, and energy budget. He ran into technical problems and also had difficulties with the Boss and D. K. regarding his own energy budget. He persevered, however, until 1968, in completing his thesis, which was not published. Sutton studied wood-louse populations on three acres of Sunday's Hill, with Kitty Southern as his adviser. Core samples of the soil-litter interface were taken in the study area, and migration was monitored by means of pitfall traps along the borders. There were seven species of Isopod present, but the millipede *Glomeris marginata* was so abundant that its biomass was greater than that of all of the Isopods combined. The tiny soil form *Trichoniscus pusillus* was at times present in numbers greater than 2,000 per square meter. He carried out a clever investigation of the predators, which included common and pygmy shrews, centipedes, spiders, and beetles, by utilizing lab rabbits to produce antibodies, for testing for isopod proteins in suspected predators (139). Another fascinating sideline involved checking the distribution of the triploid form of *Trichoniscus pusilus,* which had been recorded from Wytham woodland by John Brereton a decade earlier.

Chris Watts worked on "The ecology of woodland voles and mice with special reference to movement and population structure," under Mick Southern. His work with trapping grids using different densities of traps gave Southern opportunities to test the validity of his Great Wood estimates. He had set up the study using traps in pairs, and on many occasions both traps were occupied, casting doubt on their availability to individuals with less enthusiasm for the experience. Leslie was worried by the paucity of empty traps, when calculating the numbers of rodents present by means of the Lincoln Index. The main objective was, however, to measure long-distance movements. The movements of males were, as expected, greater in general than those of females, especially when females were in breeding condition. Almost all juvenile males and many juvenile females moved up to 400 yards from their place of birth. The effect of the food supply on breeding was also studied by supplying the animals living within an experimental grid in Marley Wood with excess food (148). The extra food appeared to extend the breeding season of the bank voles, but the possible effect on the wood mice was masked by the effect of an exceptionally good acorn crop. When comparing the foods chosen by the two rodents, Chris got rather different results from Rick Miller's, mainly due, no doubt, to the difference in the techniques used (147).

Meanwhile, Elton was attempting, through a memorandum to the Faculty Board, to dissuade them from accepting Pringle's view that the Bureau of Animal Population should cease to exist when its director reached the mandatory retirement age of sixty-seven, that is, in the year 1967. He put before them a brief history of the Bureau, and his arguments for its continued existence, hoping to get a favorable decision before he retired. After retirement, he would not have a voice in the matter.

> I consider that all the Bureau of Animal Population
> needs is to have its future settled at an early date; to be
> allowed to continue as a separate research institute in
> its present structure and form; to have its relation-
> ships—"the chain of command"—clarified; and to be
> fitted into the new science program.
>   The need for *an early decision about the future* arises for
> three reasons especially: (a) Recent confusion about

building plans for the Zoology Department, and for the Department of Zoological Field Studies, has left a number of questions in the air. The original plans arranged with Professor Hardy in 1961 were (and still would be) perfectly satisfactory as far as the B.A.P. is concerned. The next lot of plans, prepared by Professor Pringle, seemed to imply the dismemberment of the B.A.P., or perhaps more correctly, provided his answer to the possibility (that he suggested) that the B.A.P. might cease to exist after my retirement. Now that there is a real chance of rational planning of the whole thing, it seemed to me urgent that our position should be clarified before further building schemes became crystallised. (b) Uncertainty makes it harder to obtain good staff. We recently lost the chance of getting a good Senior Research Officer mainly for this reason. (c) Other administrative reasons, which need not be analysed here, but include my retirement in October 1967. My successor would need to be closely involved with the later stages of new buildings for the B.A.P.

. . . *Relation with the Department of Zoology.* Under the present Statute, this only involves our relation with the Linacre Professor as Head of the Department of Zoological Field Studies. But it has sometimes been suggested that we might again become part of the Zoology Department, a proposal which I have resisted. Ecology is a vast subject, and population ecology forms the heart of it. It will undoubtedly some day achieve a departmental level in its own right, just as biochemistry has developed in this way. Ecology by its intrinsic nature as a subject connecting living organisms with their complex environment has many contacts other than zoology e.g. botany, geography, human demography, pathology, as well as a whole range of applied field sciences including conservation. I think it would be a retrograde policy to force back into the context of general zoology an expanding subject that has already developed for fifteen years outside that Department. In fact, just as much contact is with the Hope Department as with the Zoology Department.

It was absolutely essential, he said, "that the research rooms, library, workshop and administration should all be close together as at present i.e. within a few yards or not more than one staircase away. Any plan to 'integrate' that ignores this need will lose far more than it gains. I regard the gains from centralising various research libraries and workshops as being largely illusory, and even financially very small. It would surely be a contradiction if 'integration' were to result in loss of integrity."

Elton's claim that small was not only beautiful, but was also cost-effective, was not in tune with the financial and political scene of the 1960s. In industry, big firms were swallowing small ones, and conjugating among themselves, to cut overheads and to increase return on invested capital. The British Public Service, under the control of former "captains of industry" from the Conservative party, was undergoing incessant examination for "fat," and the universities of Britain were being subjected to the same examination for "cost-effectiveness," however irrelevant that might be. Pringle was not exceptional in valuing the appearance of efficiency more than productivity. Administrators are, as Lord Snow has said, "by temperament active men. Their tendency is . . . to live in the short term, to become masters of the short-term solution." When dealing with research, questions of quality cannot be addressed as they are when producing detergents or frozen peas. Within the university, Elton's suggestion that ecology should, some day, become a full department, would not be a popular one. Some faculty members might have felt that Pringle was doing them a favor by removing competition.

In May 1963, the Board of the Faculty of Biological Sciences appointed a committee to investigate the conflicting ideas about the fate of the Bureau. Elton had been placed in the position of defendant by Pringle's attitude. He had now to make his case to a jury consisting of the Joint Committee for the Review of Activities, supplemented by two outsiders. Thus there were two college administrators: the president of Trinity and the provost of Oriel, with Sir Peter Medawar, director of the National Institute for Medical Research, and Max Nicholson, director general of the Nature Conservancy.

The committee met three times and took evidence in private from Hardy, Pringle, Lack, and E. B. Ford. No outside ecologists were asked for their views. The committee also paid a visit to the EGI and the BAP to see their facilities in the old Botanic Garden building. At the Bureau, when they inspected the workshop, D. K. lined them up and gave them a lecture on the efficiency of small specialized organizations. As they were filing out, Elton remarked to the registrar, who had accompanied the committee, that he could raise £50,000 for a new lab. The quiet reply was in the form of a question, "*What* new lab?"

There were four possibilities: (1) The BAP could be maintained as a separate research institute; (2) it could be kept as a separate institute for administrative purposes, but be physically located in the same building as the zoology department (this was the plan put forward in detail by Professor Hardy in 1958); (3) it could be kept as a separate institute but grouped physically with the Hope Department of Entomology; (4) the BAP personnel could be placed in the zoology department without any independent status. Similar options appeared to exist for the EGI, but when the legal history of its establishment was examined, the EGI proved to be amenable only to option 2.

The committee's decision to recommend option 4 was made mainly in deference to Pringle's advocacy. He was, after all, head of the department he recommended abolishing, now that Hardy had fully retired. Their reasons were published in the *University Gazette* of August 7, 1964 (pp. 1455–57):

> (a) There can be no question of the value of Mr. Elton's work as a pioneer of population ecology, and he has been admirably supported at research and technical level. The committee is convinced, however, that the Bureau suffered more and more in recent years from its isolation, which has not been purely geographical, from people working in allied fields (e.g. ecological genetics and animal behaviour in the Department of Zoology and in the Hope Department.)
> (b) The committee is in no doubt about the continuing importance of ecological studies in general, and studies in population ecology in particular, but it believes that the Bureau, as at present constituted as a small self-contained unit, and as Mr. Elton visualizes it would continue under his successor, would not be able to play a leading part in work on new problems in population ecology; and the committee is satisfied that the redeployment of resources which is now taking place, research and teaching in population ecology would benefit if it could be brought into much more intimate association with other branches of ecology than the present arrangements allow.

The report presented "the unanimous views of the committee, but owing to absence abroad Mr. Nicholson did not see this report in draft." (In private, and when it was too late to affect the outcome, Nicholson told Elton that he had not agreed with the decision, but he did not feel that putting in a minority dissenting report would have done any good.) The physicist on the committee remarked that he could see no problem in accommodating a small unit for specialized research within the new zoology department; he had a meteorological group within his department, and "having been given a corner to work in, they seemed to be quite happy."

Before any of us had seen the *Gazette*, we had been informed of the committee's conclusions in a privately printed memorandum, in the letter-to-*Nature* format, dated April 12, 1964. In this, Elton explained what the committee had been set up to consider, and the options it had been given. He told us that the fourth option had been chosen:

> As I have so far been excluded from participation in the planning of the new arrangements, I cannot give you a full picture of what they involve. But (1) they will abolish institute status and any separate social or financial existence, though they will (2) provide for about the same number of research rooms as at present, with a "group" (including visitors) working under the next Reader in Animal Ecology (3) the library will be retained as a separate sub-library unit in the Department (4) certain special services may be available, but otherwise there will be reliance on common departmental ones. In return for any loss that this involves, we are told that ecological ideas will be reinvigorated by closer contact with animal behaviour genetics, physiology, cytology, biophysics, and so on, and by a greater share in teaching. The Wytham Biological Survey may not be included or indeed continued at all, a decision about this being delayed until after the publication of my almost completed book on "The Pattern of Animal Communities" next year.
>
> . . . Some of you will know my liking for old Chinese proverbs and there are two that seem to be apposite just now. One says: *Even after a typhoon there are pears to be picked up:* and the other, a most charitable one, I think: *When a neighbour is in your fruit garden, inattention is the truest politeness.*

There was one more opportunity for Elton to promote his view that the Bureau should continue to exist; he submitted evidence to the Commission of Enquiry into the general workings of Oxford University (the

"Franks Report") (142). His case for the survival of the Bureau as independent institute was based on certain assumptions:

(a) In any university research and learning should rank equally and in their own right with teaching.
(b) Academic efficiency (stemming from the quality of both research and teaching) should not be subordinated to administrative or financial or building convenience.
(c) Academic policy is the concern of the whole University and should be debated openly.
(d) The quality of a university is enhanced by having a variety of organisations of different sizes and forms, as has long been recognized in the case of the Colleges. Scholars are not all cast in one mould. One of the advantages of a university over a Government or industrial research institute is its greater flexibility.
(e) Original ideas, like new species, may arise and flourish best in partial isolation e.g. among workers living in relatively small groups.

This memorandum to the Hebdomadal Council of the university was dated September 30, 1964, that is, nearly two months after the *University Gazette* had announced what was going to happen. He was discharging his intellectual artillery on a battlefield from which the victorious enemy had already marched away.

The decision to abolish the Bureau had been made by the time Mathews, Sutton, and Watts had completed their stay. But physically, there was little that could be changed, as the new zoology building would not be ready for occupancy for some years; in the event, it was not ready until April 1971. Elton was not due to retire until the end of September 1967, so the modest flow of students and visitors from overseas continued. Elbourn continued to work loyally on the Wytham survey and to extend Peter Larkin's work with Fager's artificial branches. A staff member of the Canadian Forest Service, C. H. Buckner, came in 1965 for a year. He had done ingenious work on the Larch Saw fly and its predators, and he now turned his expertise to looking at the role of shrews as predators of George Varley's oak moths. Varley had suspected that shrews were a major predator ever since 1950, when he caught them in his insect traps. Buckner now showed that they were important in reducing the numbers of oak moth pupae surviving the winter in the soil (8). Another visitor for a year was R. S. Peterson, an expert on seals from

Johns Hopkins University. He did a valuable job of building up the BAP library's coverage of seal biology. Another Rhodes scholar arrived in 1965 to complete his D.Phil. thesis under Elton's supervision. This was Dave Schindler, who had come from North Dakota earlier to study fish behavior under Niko Tinbergen. The facilities for studying fishes in the zoology department proved to be inadequate for the kind of work he wanted to do, so he had changed over to a study of *Daphnia* under Peter Brunet. He had worked out the carbon budget of the little pond crustacean using C^{14}, and used the results to predict the population dynamics of a population in a small pond near Oxford. This project was ill-fated too; a road-repair crew wiped out the *Daphnia* by contaminating the pond with oil. Then he started a study of freshwater plankton and nutrient cycling.

While coping with these complications, Dave got to know the Boss through attending seminars at the Bureau. Elton was sufficiently impressed with him to hand him a front-door key, to enable him to use the Bureau library at odd hours when the university libraries were closed. After a year back in the U.S., Dave returned to Oxford to write up some work he had carried out on a small lake in Minnesota. It was now clear that his thesis was going to be more about ecology than physiology, so Brunet and Elton arranged for him to switch supervisors. Dave has commented: "I got along well with Elton, despite what I'd heard about him being hard to get to know, and I was continually surprised by the insights he had into the workings of plankton populations (although to this day I'll bet he won't admit it)." He was right about Elton's reluctance to claim any credit; the Boss had already told me that Dave's thesis was first-class, although he couldn't follow most of it.

The last woman to undertake research in the BAP was an Oxford graduate, Dianne Love. She began, in 1965, to work on small mammals in the Pasticks, with Southern as supervisor. She married Peter Evans of the EGI and later wrote up her work for publication (62). Two other graduates started their projects in the fall of 1966, knowing that they would be transferred to the new Animal Ecology Research Group before their work was completed. The transition would be smooth, as the AERG would stay in the Botanic Garden building for some time, and the Southerns and D. K. would also be working in the new setup. These two young men were of a caliber that ensured the Bureau's research quality would be maintained until the end.

R. L. Kitching came from Imperial College, sponsored by the Nature Conservancy. He accepted a project, dear to the Boss's heart, of studying the animals living in tree-holes in Wytham (83). According to Elton, he did "excellent incisive work that put this bit of our ecosystem well and truly on the map." Kitching handsomely thanked D. K. for his help in designing a new core-sampler for semiliquid substrates (82). Elton's last student, like his first, was an American Rhodes scholar. Jim Schindler, like his older brother Dave, was a limnologist. He undertook the BAP's first real lake study, the ecology of zooplankton populations in the reservoir at Farmoor, just outside Oxford, near Wytham. His thesis was a complex one, with much quantitative work on the algae and water chemistry as well as on the fauna. To some extent it was slanted to fit in with the philosophy of the AERG which became, under John Phillipson, strongly oriented towards "trophic-dynamic" aspects of ecology. But it also reflected a developing interest in the possibility of "defining aquatic pelagic habitats using thermodynamic principles." This was to dominate Jim Schindler's research for a decade after leaving Oxford. Jim wrote, "It works and it's fun, but what is the strangest facet of all is that I derive many of my precedents from Elton's material. His work wasn't that quantitative, but it had a solid logical basis coupled with his way of expressing ideas and concepts to give a vital feeling to what he wrote."

Charles Elton was fortunate in that for most of his life he worked under able and generous men, who did not feel threatened by his development of a new kind of zoology. Of Professor Goodrich, Elton remarked that his was an unexpected and generous attitude. Although he had very definite ideas about what properly constituted zoology, he gave Elton a job which would enable him to propagate quite a novel view of the subject. He also backed him through the difficulties raised by others of less flexible views. As well as being the greatest comparative anatomist in the world, he was a great man. After Goodrich, things could have changed for the worse. J. Z. Young was so certain he was going to become Linacre Professor, he had told Percy Trotman, the departmental technician, that his first act was going to be to fire him. Percy related this to me with great relish, as he was still in the department in 1980 at age seventy-seven, having been called back from retirement. (A few years later, the university honored Percy by awarding him the M.A. degree.) Fortunately for him, Alister Hardy had got the Linacre professorship, and he had proved to be a gentleman in the best British tradition.

There is no room for doubt that Hardy's successor, Professor J. W. S. Pringle set out to destroy both the Bureau of Animal Population and the Edward Grey Institute for Field Ornithology. There seems to be no doubt either, that he did so because he felt it was the right thing to do; he felt

Ken Marsland, Denys Kempson's assistant (and, later, successor), with Jim Schindler, sampling plankton in Farmoor Reservoir, near Oxford. 1967. Courtesy of Charles Elton.

he had a mission to cut off these two warts which had grown on the body of the zoology department. He brought with him a reputation for getting his own way; there was even a joke about the verb "to pringle," and some spoke of the Bureau having been "gepringled." Pringle couldn't understand why Elton was upset when he suggested, as a first step in streamlining the administration of his department, that Elton should report to Lack. Professor Blackman, of the Oxford Department of Agriculture, explained "that's because you don't understand the Oxford *ethos.*"

Being small and separate had great advantages for a research unit existing in a benign environment, but those same attributes were of poor survival value in a hostile one. The Bureau's kind of research was specialized; Pringle could describe it as "awfully narrow." The heart of the library, and its most valuable research tool, was the reprint collection; Pringle questioned the need for a separate library which was "mainly a collection of reprints." In that context, the word "reprint" sounds less valuable than the original publication. For twenty years, the Bureau had sought to function as an international institute, and the presence of its

alumni in important jobs worldwide was evidence that it succeeded;
Pringle could make the comment that it had catered to "so few U.K.
graduates." Elton had been trained to box under the constraints of the
Marquess of Queensbury's rules. That had not prepared him either for
barroom or boardroom brawls, in which very different tactics are necessary. He would have given a good account of himself in the former situation, once sufficiently aroused to remove his glasses, but he was far too
straightforward in his dealings with people to be effective in the power
games of a British boardroom. Faced with Pringle's inexplicable hostility,
masterly use of lobbying and innuendo, and the sheeplike behavior of
many of his fellow faculty members, he was very frustrated and angry.
Having kept out of university politicking, he lacked alliances and prom-
issory notes he could call in. He was under a further constraint, which
may have given people the impression he was not doing enough to fight,
and that was the obligation of the "Boss" to look cheerful, even when
he might be losing. Having stayed off the academic battlefield, he lacked
knowledge of the terrain over which he was forced to retreat. Nevertheless, he gave a good account of himself, and was later told by a friend on
the Board of Faculties, that Pringle won only by a very narrow margin.
And Pringle had to concede the continued separate teaching courses and
research in ecology by setting up the AERG. Elton also succeeded in
saving the integrity of the BAP library, and got breathing space in which
to maintain the Wytham Biological Survey. When *The Pattern of Animal
Communities* was published, it became clear that Pringle knew as little
about ecology as Elton knew about molecular biology. In his book Elton
had stressed the necessity of preserving voucher specimens (identified
with certainty by an expert in the group) and the notes which accompanied them. Pringle's comment was, "Now you'll be able to throw away
all of those notes and specimens."

The comments which flowed back to Elton in response to his note
about the closing down of the BAP were, as might be expected of those
who responded, supportive of his point of view. A few urged him to
adopt a posture of Christian resignation. But indignation was expressed
by the majority.

Andrewartha and Birch, whose massive textbook had received less
than high praise in Elton's review, held no academic grudges, and sent
sympathetic and supportive letters. Andrewartha, like most others, had
assumed that the Bureau, which had become so well known all over the
world, would have been allowed to continue to exist. It really was a
shame, he wrote, that the Bureau's "mellow fruitfulness" had to be sac-

rificed to the fetish of administrative tidiness. Charlie Birch expressed the view that no other ecological institution had influenced so many people, and that their work and thoughts continued to reflect its effects.

Maurice Solomon wrote from England:

> This is depressing news. The inner workings of Oxford University are of course unfathomable to outsiders. I can only say that, to animal ecologists here and abroad, who know what you and the Bureau have done towards founding their subject and stimulating its development, the closing of the Bureau (without plans for equivalent new developments) will seem a destructive act, a great setback . . .

David Pimental wrote from Ithaca, New York:

> In spite of the almost unbelievable news that the B.A.P. is to go into oblivion as an organization, its long years of solid and worthwhile endeavor under your leadership will stand the test of time. This impetuous type of action is attacking the biological communities at other major universities. Already in the States many are involved in the Oxford type of reorganization, that is, consolidating and coordinating the "life" out of biology. Here the molecular biologists quickly gain the upper hand in the power struggle since this seems to be the most popular subject. Taxonomy and ecology, in many cases, have been forced out of existence.

Sir Douglas Veale, who as registrar of Oxford University had been most helpful to Elton, viewed the destruction of the Bureau as just one facet of the general trend towards administrative conformity and rigidity. He was horrified, he wrote, to see small independent units being swallowed up, and Oxford becoming a place of large units, which would stifle the personalities of the "helots" who would work in them. No doubt he would have been even more horrified to know that the same "streamlining" of university departments was taking place in the United States.

But these sympathetic letters from all over the world, while helping Elton to live with the situation, could have no effect upon events. The deed was done, and it only remained to wait until September 1967, Elton's mandatory retirement date, to bury the body. As far as Pringle, a molecular biologist, was concerned, the Bureau had already ceased to exist, and he did not bother to give it a mention in his departmental reports after September, 1965.

Elton was thoroughly wedded to fieldwork in Wytham, and while on brief annual vacations, to making observations in other parts of Britain. He also had experience of Arctic environments. But he had never worked in tropical habitats. At this traumatic time in his life, Harold Trapido, in an act of great kindness and superb timing, arranged for a two-month fellowship to put Elton in to a stimulating environment— the neotropical forest of Central America. His initiation into these new mysteries was followed by more visits financed by the Smithsonian's award of a Browning Fellowship. Taking his Wytham techniques to the tropics brought home to Elton the rarity of individuals of any one species, and the diversity of species to be found in tropical habitats (53). The Eltons traveled via Washington D.C. at the time of the Martin Luther King marches. There were riots and fires, and National Guardsmen in battle helmets everywhere. The Boss was delighted (it was reported to me), and said that he had always known America was like this! This was a tongue-in-cheek remark, of course, as Elton had visited the States before. But he had experienced a brush with law enforcement during his third visit in 1938. Tracy Storer took him to look at the classic life-zones in Yosemite, and they visited Madera. Some counterfeit money had just been passed at the general store, and the sheriff "standing about seven feet tall and carrying a six-gun" spotted Elton as an obvious stranger in town, and started to question him. Now Elton has a habit (Mick Southern caught it from him) of hitching up his pants when he is nervous. His hands now went down to his belt, and the quick-acting sheriff pinned his wrists in huge hands and started to take him to the lockup. On the way, they met Storer, and the sheriff grabbed him too. Fortunately, going through town they came across the local head of the state Wildlife Department playing cards on the sidewalk with cronies, and he identified the two desperadoes. The contrast with British-style policemen would have made quite a mark.

There is nothing objective about my attitude to the destruction of the Bureau, and most of those who heard from the Boss probably shared my disgust, not so much at the decision, but at the way it was forced through some years before Elton's retirement. But there must have been room for another point of view, and some support for it within the faculty, or Pringle would not have got away with it. Most of the people who worked in the Bureau found its ambience to their liking, and recorded their feelings at the end of their thesis. David Jenkins's thanks was, perhaps, the most elegantly expressed: "I am greatly indebted to Mr. Charles Elton, F.R.S., for all the facilities offered by the Bureau of Animal Population, and for his tolerance towards a research student who was sometimes not seen at the laboratory for weeks on end. Such absences, though enforced, were to the disadvantage of the student, since the friendly and cooperative atmosphere of the Bureau, under Mr. Elton's guidance, proved stimulating to a remarkable extent."

There were two who simply thanked Elton in the minimal obligatory fashion, and they were both from the same overseas department. The casual atmosphere did not suit them, as they had expected to be instructed and directed with the more usual professorial authority. This is what one of them had to say when I asked him what he felt being at the Bureau had done for him: "Vaguely I have inherited Charles' disrespect for statistics, but I think this was a reinforcement of my own inclination and was, in retrospect, a bad thing. Also a sort of feeling that the answer will come if you wait long enough or think about it long enough, rather than through active experimentation. Certainly there was no air of frantic research and precious little sense of overall direction when I was there. Rather an air of quiet wonder at nature that was sustained by random observations on anything that moved or grew in Wytham. As a philosophy of life this seems to me to be ideal, and I hope I have absorbed some of it. As a grounding for what is now a highly competitive profession, it was inadequate." When this student was at the Bureau it had no future. The direction of research in the years ahead had yet to be determined, and all current projects were being wound down, so it may well have been true that a sense of direction was lacking. A decade earlier, it was very much in evidence, according to the comments of another former student: "In just one year in the Bureau I think I became sold on the idea of patient, persistent, long-term research in the ecological sciences. I could see the projects of Charles Elton, Mick Southern, and Dennis Chitty which were all based on very thorough thought and long-term data collection to probe basic questions in ecology. There was not so much flitting about from one project to another as we see in this country; not so much chasing popular rainbows of the moment; and not so much effort at grantsmanship to do the popular topic of the day. I think this is now what plagues American science. Under our political system or research support one is under continual pressure to produce applied results . . . with "war on rats" one year, "war on cancer" the next, and "war on Alcoholism" the next. Perhaps we as a Nation were overimpressed with the Manhattan Project, and this initiated the habit of crash programs. Anyway, it's hard if not impossible to get long-term projects supported with any kind of stability, so I look back on the Bureau as a period in the history of science when a small group concentrated on doing certain basic things very well and very thoroughly."

These sentiments pleased me greatly, as they reinforced my own, but to my surprise and discomfort, he continued, "The only academic problems which worried me, I think were the barriers between Elton, Tinbergen and Lack. Here were three excellent groups all very territorial and walled off from one another, and this seemed a great shame. I think the students got around this without too much trouble—we had open access to Tinbergen's group and in fact, went to his home every Friday night, and we also went in and out of the Edward Grey as readily as we liked, but I recall almost no contact between the staff and faculty."

Obviously, opinions differed about the advantages of having separate institutes, and we are bound to consider, with the wisdom of hindsight, what would have happened to the Bureau if its independence had been maintained. Could a Bureau of Animal Population have been a first-rate unit without Charles Elton? We have to consider his ideas about how a small research institution ought to function, as he was one of the few zoologists who have created and directed one. It functioned very well because he was in charge, but that does not invalidate his general ideas. On the contrary, it may be, that without him, it might have functioned, for the first time, as he considered it ought to function! In order to justify its continued existence it is not necessary to demonstrate a need; it already was a part of the university and a modest item in its budget. It is not necessary either to suppose that it should enjoy such eminence in the rest of the world; obviously much of that was due to Elton's personal genius. It is only necessary to show that it would have continued to provide an environment for first-class research and instruction.

Elton's memorandum to the Franks Commission of Enquiry included the following:

> Ecological research, on account of its lengthy and complex character, requires certain special conditions for its work that differ from those of other branches of biology, and the research institute provides an ideal environment for it. Besides being in a state of rapid expansion as a fundamental science, ecology is of extreme urgency in the practical affairs of the world. . . . It is no longer just a growing point from the general base of zoology, but a discipline in its own right that extends far outside

the study of animals alone. Indeed it is one of the major features of biological advance recently that the field of zoology has patently split into two separate ones that can be called physiological and ecological, with certain subjects like genetics and behaviour partly bridging them. Ecologists study not only animals but the whole of their environment—other animals, plants, micro-organisms, soil, water, climate and the influence of man. To push ecology back into the framework of a Zoology Department is about as justifiable academically as it would be to push biochemistry back into organic chemistry, physiology back into human anatomy, or forestry and agriculture back into botany (142).

As for the working environment, Elton's ideas had already been given in his talk at St. Hugh's in 1947:

No ecologist can hope to experience the "deep, deep peace of the specialist"; but he (or she) should be able to enjoy the work in spite of its formidable scope and the rather austere sampling difficulties that confront him. We find here, I think, a force which will cause workers always to gravitate to some extent into voluntary working teams, in which the strain of field operations can be shared, as well as new ideas. The great thing is that the associations into groups should be voluntary and not laid down from above. Unless research workers keep the utmost freedom to try unpromising-looking ideas, if necessary for a year or more, and to do "fool experiments" (as long as they aren't too expensive), and break away from the hardening traditions of the group they are in, (the) place will not stay alive. In other words, the real job of a Director is to get people the opportunities and facilities for satisfactory work, and then make sure that they remain intellectually alive in it. The library, office and laboratory staff have the job of seeing that these opportunities are smoothly arranged.

The most senior member of the Bureau's staff, at the time of its dissolution, could have maintained that environment, if he had been given the opportunity. He was not only a research ecologist of vast experience, he was also (what the majority of ecologists are not) a naturalist. When the teaching of the zoology department had been reorganized in 1961

Two founding members and elder statesmen of the Mammal Society, at the 1965 annual meeting. Courtesy of Charles Elton.

and 1962, he had been brought into undergraduate teaching. As for experience (and successful guidance) in graduate work, Southern had more experience than anyone else in the department. Southern's most important attribute in succeeding Elton would have been his similar attitude to the philosophy of ecological research. He too did not accept "the model of the controlled, pre-designed, true-or-false experiment of the physicist or chemist—a forcing of an answer from nature," and would agree "that another model is possible and indeed, does exist in science—the receptive, descriptive, let-be observer whose aim is to perceive fully and to understand fully" (103). But that very qualification was a liability in an exercise designed to eliminate administrative heresy.

Ever since its founding, in Birmingham in 1954, the British Mammal Society had been guided by Southern as Elton's surrogate. His weakest area would have been the invertebrate work of the Wytham Biological Survey. But by the year of Elton's retirement, the mainstay of the survey work was Kitty Southern. As a team, the Southerns and D. K. and Ken Marsland, could have carried on the quality and traditions of the Bureau's research and teaching. Elton would have been available as an elder statesman, and his diffident manner would have ensured that he did not interfere in the day-to-day running of the institute. But continuity was not wanted. Southern applied for the readership in the new Animal Ecology Research Unit and was interviewed of course. At the interview he was greeted by Pringle with the remark, "This is most awkward for both of us." Mick felt that was distinctly odd, unless the decision had already been made that he was not to be appointed. The position went to John Phillipson, a senior lecturer at Durham University, a modest man who "would not have dreamed of applying for the job if not encouraged to do so by David Lack and George Varley." Phillipson's sponsors were both, like Pringle, Cambridge men.

The Edward Grey Institute remained intact, thanks to the legal provisions of the original bequest to the library, and the Hope Department of Entomology retained its independence until Varley retired. The Bureau library was not disintegrated and diffused into the stacks of the main departmental library, but kept its integrity in a section separated by a folding grill. It is called the Elton Library. The marvelous reprint collection has continued to grow under the loving care of Mrs. Dunkley ("Mrs.D"), and in recent years, of Mrs. Diana Thomas.

Elton retired on September 30, 1967. John Phillipson had been appointed in May of that year, but he did not arrive until a month after Elton had vacated his office. John made two physical changes right away. He felt uncomfortable about occupying the Boss's center of activity within such a well-established territory, so he put down a carpet in Elton's old office and made it a staff tearoom. And he unlocked the door in the corridor upstairs, where it joined the EGI.

8 The Bureau Experience

In 1978 Elton commented, "I would only object to your book if it overemphasized me in the BAP picture, or tried to be biographical in a larger or more general sense." Two years later, knowing that I had not given up the project, he again expressed his apprehension about being cast in a starring role. He felt, he wrote, that a book was in order about "the life and times and accomplishments of the BAP, not omitting anything short of libel about its demise." The draft synopsis I had sent him made him feel that he would not "hog too much limelight." It was important, he emphasized, to make it clear that although he played a prominent part, "the whole nature of the BAP performance was its mixture of team-work and freedom," with him adding opportunities and ideas. In 1989 I still find it is not possible to give an honest account of how the Bureau functioned without risking betrayal of his trust. But one cannot produce *Hamlet* without the Prince of Denmark. The few anecdotes I record seem to me to be of historical rather than personal relevance.

The all-pervading influence of the Boss was his most striking characteristic, in spite of his diffident manner and reluctance ever to give the *appearance* of being in charge. I don't think any of us ever planned a piece of work or wrote up our results without having the conscious or subconscious concern that our efforts might disappoint him. I was fortunate in that he never expected very much from me intellectually, and I was sufficiently industrious to earn his approval. His extraordinary desire to avoid having his name at the top of the playbill persisted into retirement. When the *Sunday Times* ran a series of features on the men and women they considered to be "Makers of the Twentieth Century" they included photographs of all except Charles Elton. As he refused to submit a photograph for publication, they sent a photographer to Oxford, and he hid in the shrubbery of the traffic island outside Elton's home in Park Town, in order to get a picture as Elton came out of his front door. But the Boss came upon him from behind, and said quietly, "We don't like people trampling our shrubbery. Do be a good chap and go away." So the column on Charles Elton was the only one printed without a portrait of the subject. Instead, a medallion bearing the head of Linnaeus was used.

Charles Elton has consistently sought to play down his leadership role. He happened to be the member of the team who was in charge. Making decisions about the direction of research by a democratic process had worked well during the war. In one of his rare speeches, this one given to mark the setting up of the Bureau in its quarters at 91 Banbury Road, in April 1947, he remarked:

> During the war we solved some of our problems by individual originality, but many others by the cut and thrust of discussion and the operation of various kinds of "multiple brain," such as our discussion panels upon reports to the Government. Please note that although the worker was always given the final say on any controversial point, there was no single instance during the war of anyone reaching other than an agreed solution. Furthermore, no serious mistake was ever made in giving advice through these reports, of which there were over 100, and which did not lack originality or boldness of treatment on that account. In the last several years the leadership of research was fairly evenly spread among the seniors, although the Director was occasionally allowed to exert himself on general policy. The chief dangers were that younger research workers would never learn thoroughly to stand on their own feet, even under such a modified system of "directed research," and that eventually we should all get too set in our ideas of, it might be, "cycles" or "prebaiting" or "age determination."
>
> Therefore, since the war I have left each worker to follow and establish their own line. I take the view that the magnetic draw of the subject we are in, and the special training and outlook people get, are sufficient to prevent the lines of our research drifting quite widely off the mark. Let me make this clear; if someone decides to study fish populations here instead of mammals, he must be clear whether his training and the facilities here are adequate for the job. The principles may be the same as with mammal populations. On the whole, I think it will pay most of us to stick to mammals and their enemies (whether mammal or bird) and perhaps their parasites, as material for study, because we shall in

that way build up a more effectively trained group. I am not going to do so myself, as it happens. But if someone decides he wants to study the physiology of the nervous system or the cuticle of insects, let him in all reason be prepared to seek a job elsewhere, where his talents may be best utilized. Just because we want to stay a small research group, we cannot scatter our interests to the same extent as a big department. Somewhere between overspecialization and overspreading of interests is the zone we want to stay in, and I hope to keep a flow of fresh interests and points of view through research workers visiting here."

The main reason almost all of the students admired Elton was that he had a rare combination of genius and modesty allied with great self-control. Rick Miller has summed it up: "He has this quality . . . of acting as though you know about anything he mentions, when in fact he must know that you do not. The result is that he leads you into theoretical discussions where you are sometimes out of your depth, and then treats your ideas and opinions as equally valid. It's a way of drawing your insights into problems, and your ability to use your mind, rather than on factual knowledge per se. I've never felt that I was his intellectual equal, but I've never had to feel that my opinions were less important." When John Brereton was asked for his comments twenty years after his time in the Bureau, he wrote: "I am finding that as time passes, my respect for the Bureau grows; I left feeling that I had not gained much at all. Now I find Charles' stature getting bigger all the time—and the whole association seems like a long encounter group whose impact becomes more telling—but I am slow to learn. I'll write more later." But like all of us, John was far too busy to write letters. When I wrote a year later, he replied that he was just off to the Sorbonne for a year, and would take my letter with him and reply at leisure. But soon after arriving in Europe, John learned that he was to have no leisure time—he was dying rapidly of cancer. So he returned to Australia, and found time to call and say "Goodbye" to the Boss on the way home.

Elton's remarkable self-control may be, in part, the result of training as a boxer. As a boy he was trained by Jack Houghton, an ironworker, and Jack used to poke him on the nose again and again, saying all the while, "Now don't lose your temper, Mr. Charles, don't lose your temper." While Elton boxed for Oxford against the Army, he did not go against the Cambridge team, so he did not become a boxing "Blue." He gave it up before his final Schools, but it made him very adept, he said, at dodging shopping baskets in the crowded Cornmarket. It was also useful in Spitsbergen, where he had to camp in a large packing case for some weeks, and shadow-boxed to keep himself warm.

Students at the Bureau had, naturally, a tendency to be earnest, but the atmosphere was lightened by Elton's sense of fun. At the same time, he was capable of exerting, mainly by his example, the discipline that is necessary in research, or for that matter, in any difficult enterprise. Although we always referred to him, out of hearing as "The Boss," with D. K. sticking to his own use of "The Chief," there was no overt ritual for the recognition of his authority. He played the part of "The Boss" very seldom, so when he did, its effect was devastating. Oxford is a very seductive place, especially for people from overseas, as it has an extraordinary richness of extramural activities. If outside interests seemed to be taking too much time, or causing *slacking*, a quiet word from the Boss was enough to bring anyone back to work. For example, when one graduate student met him in the corridor, while making an exit with an armful of hockey sticks, Elton nodded at the gear and said gently, "You know, you'll soon have to make a decision—a very important one in your life—whether you came to Oxford to work or to play hockey."

I always imagined the Boss to be surrounded by an energy shield when he did not want to be approached. He would move swiftly to a shelf in the library, pull down a book, deftly turn to a page, check some detail, put the book back, and pad back to his room. Even if one wanted to consult him about something, his purposeful manner signaled that he did not want his thought diverted. I fancied that to interrupt would not only be rude, it might be dangerous! He has always been reluctant to admit that he is intimidating. After his retirement, he remarked with regret that the students in the new Animal Ecology Research Group seemed to avoid making use of the Wytham survey material even when it could save them work. The material was stored in the room set aside for Elton's use and there was seldom a knock on his door. I was frank enough to suggest that they might be shy of consulting him not merely because of his seniority and eminence, but because they might find him hard to approach. "You can be pretty formidable at times," I said. He made no answer, but my remark must have made an impact, for when I saw him the next day, he greeted me with: "I've been thinking about what you said. Formidable, you said. Well I think you can be pretty formidable at times too!"

Although intolerant of slackness because the resources of the BAP were so limited (he had no interest in criticizing lazy people elsewhere), the Boss used to warn newcomers not to take on too heavy a program of routine work. "When you plan your work," he told me, "be careful to give yourself time to *think*." In his last book (p.169), he writes of the complexities of dune communities:

> Perhaps the only way to become seized with the meaning of all this is to abandon books for a time and sit or wander slowly on a sea cliff slope or rolling dune on a day of good weather watching, listening and imagining for many hours. I get the impression that this is one of the things ecologists too seldom allow themselves to do . . . (51)

A consequence of undertaking regular fieldwork in Wytham was that it did give students some time to think about what they were doing. Too much time in the field, for long periods, can lead to miserable loneliness, but Wytham was only twenty minutes from downtown Oxford, and it was usually possible to plan fieldwork so that we could be back in the Bureau for tea. If the Boss was not there, conversation at teatime might be interesting. If he was present, it was sure to be. Richard Miller has described why he enjoyed his BAP years:

> The senior scientists each had different qualities and formed an incredibly helpful and stimulating mix. Dennis' hard-nosed scientific idealism and standards; Mick's gift of first-class natural history and fund of knowledge; George Leslie's patient counsel in matters mathematical, and his unique balance between theory and practicality; and of course an orchestration of ideas by Charles. . . . And each of these people, and the students, always had time to listen to the other person's ideas or problems, and to provide advice and criticism. There was more interaction and knowledge of each other's work than I have seen anywhere else, before or since. And of course, the staff of D. K., Marie, and Marjorie always were interested in what you needed and provided continual support. . . . One important fact is that your ideas were always challenged and listened to. It was not an intellectually dull or complacent atmosphere. If tea time conversation began to be commonplace, Charles could always be counted upon to say, "Richard" with a rising inflection, "What do you think

about the fact that . . . ?" which of course was almost always a fact that you had never heard of. Or, "Peter, what is your theory on . . . ?" We learned a lot of ecology this way, and through our knowledge of each other's research.

The working atmosphere so accurately described by Rick Miller was not merely a phase of the early fifties. It was maintained with remarkable constancy for fifteen years. Dave Schindler, another American Rhodes scholar who worked in the BAP in the sixties, and went on to direct his own small institute commented:

> The B.A.P. had that sort of monastery-like quality that closes out the outside world and all its frivolities, to allow one to concentrate exclusively on science while within its doors. It's a characteristic which I've attempted to copy exactly . . . with some success. We're even about the same size as the B.A.P. An optimum size for a research unit is just large enough to survive without an *in situ* administrative staff. That's just one too many trophic levels for a scientific food web. I still haven't seen a larger unit that's been able to maintain a high quality research program for more than a couple of years. . . . Curiously, my outlook has gradually become more "Eltonian" as I get older . . . it takes a stimulating environment to still be having an effect 17 years later.

The very last Bureau "graduate," yet another American Rhodes scholar, wrote to me in 1983 how he had gone to Oxford expecting to find a modern laboratory with the latest gadgets, and that he had expected Elton to be a typical big-name scientist, "cold, aristocratic and arrogant!" Instead, he had found both the Bureau and the Boss to be "small, warm, old, and a little worn."

> Charles met me with the warmest handshake I have ever received from a big-name scientist. . . . He wore a simple sweater and tie with old woollen trousers— nothing fancy, but ever so functional . . . . he had the bluest eyes I've ever seen. I expected that research at the BAP would be like American research—"one-upmanship"—high technology (but often lowbrow). It took exactly two days to get it all straight. Elton told me how things were done at the BAP . . . . it was cerebral

science—not technological science. Elton used money for books and journals, not the toys of technology. He also told me to attend afternoon tea—research was discussed during tea; ideas, procedures, results, whatever. I also discovered the importance of technological assistance in research at the BAP. It had D. K. and Ken to do most of the design and development of instrumentation. This was new to me. Most American universities have "shops," but they aren't too helpful for research and design. Also you must pay for all of the time and material that's used in equipment development. The BAP's shop was a permanent component of the research establishment . . . . D. K. and Ken were incredible people . . . professionals in every way and proud of their work. They were also completely dedicated to Elton and the BAP. D. K. used to claim that he could build anything—but I'd better be sure I needed the equipment. He demanded that any work done be justified in terms of the research requirements—it was the only way to make efficient use of his time.

Denys Kempson was indeed a remarkable man. He belonged to a generation of Britons who had been denied tertiary education by barriers of class and money. He was professorial material, but he entered academic life through the backdoor—the workshop. D. K. worked with, rather than for, men of the highest calibre; Julian Huxley, J. B. S. Haldane, J. S. Watson, John Baker, and Charles Elton, and the association was fruitful of ideas and inventions (76,77). He scrupulously kept his private and professional lives apart. His wife had little knowledge of what he did at work, or even within his home workshop. When he died, suddenly of a stroke, in February, 1977, the Boss wrote:

I have lost one of the most interesting, loyal and close friends I ever had, as we had worked together since the late 1940s. He did a quite unbelievable job, with standards we all know about. I think I especially learned from him how to be more careful in looking at a job and planning it, before starting to do anything. His deliberation was impressive. And, as Rick noted in a letter recently, D. K. would start by giving one a short lecture, and then show that he knew more about scientific method than one did oneself. . . . I remember one superb evening, when D. K., Tom Park, Dennis, Mick and

> I were drinking beer in the Lake Vyrnwy Hotel, and
> D. K. in a relaxed state suddenly remarked: "You
> people only investigate, but I create!"

Thomas Park has told me that this occurred on November 25, 1948,
when the party were sitting around a handsome fire and sipping a rare
Scotch whiskey, and he requested all of them to write a memorial statement about a splendid day in the field. Helen Chitty supplied: "What a
show the 'inspectors' put on! We didn't stand a chance." Dennis commented: "We hate to have our evenings' work interfered with as a rule
but this occasion was an exception." Mick Southern, as the other vole-
catcher present, rose to the occasion with:

> Oh thou, who didst with vole-trap and with *Gin*
> Besit the path we were to wander in,
> Thou will not with the Lincoln index record
> Our lives (tables) and impute our fall to *sin.*

Elton's contribution was: "Here it is! May your theories never be de-
throned by a New Avalanche of facts."
     Although highly imaginative in his work, D.K. tended to triteness and
repetition in his remarks. On innumerable occasions he glanced at the
clock in the workshop and said brightly, "How's the enemy?" and when
lovingly stroking the surfaces of a piece of metal he was about to install
with one face exposed to view, he would announce his decision with,
"Best road to London, eh?" So his remark at Lake Vyrnwy was a rare
flash of revelation of the private man he kept apart from the technician.
Although he maintained a professional relationship with the visitors,
D.K. was a generous and sympathetic listener when one got into personal trouble, just as the Boss was.
     Amyan Macfadyen was also an inventive gadgeteer, but with a very
different style (78). After D.K.'s death Amyan wrote:

> Apart from the strictly practical side, D.K. was the only
> person in the BAP for most of the time I was there with
> whom I could discuss technical (electrical, electronic,
> and mechanical) matters, and even if it was a field at
> the edge of his competence he was always willing and
> eager to consider new ideas and offer helpful criticism.
> I certainly owe him a great deal for his companionship,
> especially at the beginning when I felt pretty over-
> whelmed by the intellectual stature of the Boss and
> other senior people.

Charles Elton, Tom Park, Dennis Chitty, and Mick Southern in the field at Lake Vyrnwy, Wales. 1948. Photograph by Denys Kempson. Courtesy of Charles Elton.

For his part, D.K. was horrified by the untidy tangle of wires and makeshift equipment with which Amyan contrived to surround himself, and through which one had to cautiously pick one's path. When Amyan had an electrical fire in his room, D.K. said solemnly, "I'm afraid I shall have to tell the Chief about this." and Amyan replied, "Oh D.K. that's quite all right. I'm much less afraid of the Boss than I am of you!" Although this was amusing, Elton was not amused at the prospect of a fire destroying the library and his material and records. This incident probably contributed to the decision to suggest to Amyan that he advance his career elsewhere.

Marie Gibbs, although she kept a low profile, was just as important in maintaining the unique Bureau atmosphere as was D.K. After working in a bank for four years, Marie worked for a local authority (the British equivalent of a county board in the U.S.). Then she learned about a vacancy in the Bureau through a friend of hers who shared digs with the young lady who was giving up the job. She bought a new hat and borrowed a fur coat in order to appear well-groomed at the interview, and then found Elton in an old pullover and tennis shoes. "We were *both* terribly embarrassed," she recalls. She got the job, and supposes it may have been because of her experience in a bank. Elton wanted a secretary

Two staunch loyalists of the BAP: Marie Gibbs (secretary) and Marjorie Nicholls (former librarian), at Marie's retirement party. 1967. Photograph by Denys Kempson. Courtesy of Charles Elton.

who could add up columns of figures accurately; he was fed up with the mistakes frequently made by her predecessor. He suspected that she added in the dates at the tops of the columns of the snowshoe rabbit returns. Marie's role grew in importance until she became what Elton pleasantly described as the Bureau's "Private Relations Officer." She made one's move to Oxford easy, by getting to know lots of landladies and picking temporary digs which matched one's means. Then when people got settled, they could look around at leisure for something more permanent. As Rick Miller summed it up: "Marie did a damn sight more than put out tea and biscuits."

Marie joined the BAP on March 1, 1935, and retired on the same day as the Boss, September 30, 1967. One day in 1980, while having tea with her, and talking about the difficulty I was having in writing a history of the Bureau without upsetting the Boss, I said, "You know Marie, he couldn't have done it half so well without you and D.K." At this she looked quite alarmed, and replied, "Ooh Pete. He wouldn't like to hear you say *that!*"

No formal courses were conducted within the Bureau. The doctoral candidates at Oxford (and Cambridge), unlike those at most lesser universities, were not required to take courses or to qualify in subjects other

Cartoon by Ken Marsland to honor Marie Gibbs on her retirement. Courtesy of Ken Marsland.

than their research topic and relevant material. Most long-term visitors attended Elton's lectures for undergraduates, and some, including myself, jumped at the opportunity to attend the superb lectures of Niko Tinbergen, Oxford's first lecturer in ethology. A few also attempted to stay the course of Dr. D. J. Finney's lectures on the statistical design of experiments. But our statistical training tended to be absorbed in the same informal way as our ecology. This was due to the benign and pervasive influence of George Leslie.

Leslie had suffered from lung illness while a student, and took care not to overexert himself physically. He worked mostly in his home in the afternoons, so he was not a regular participant in afternoon tea discussions. But he was generally in the Bureau in the morning, and was available at the "standing" morning teas, which were not in the library, and at other times, for friendly counsel and discussion. He did not push his services or his advice upon anyone. One was just advised by Charles or Mick or Dennis: "Better have a chat with George before you go any further." He would listen to what you had to say, gently probe to clarify your objectives, and then say that he would think about it for a day or two. At the next meeting he would suggest, with the greatest courtesy,

Professor Sir Alister Hardy and Richard Freeman at Marie Gibbs's retirement party. 1967. Photograph by Denys Kempson. Courtesy of Charles Elton.

how to go about making good the unfortunate (he was too polite to say "dumb") gaps in the material, and what were the appropriate tests to apply. One learned to talk with him *before* gathering information, rather than relying on him to bandage up weak material. According to Elton, Leslie was "the real intellectual of the BAP, without whom it would not have risen to such high esteem." Whether or not that is going too far, it is certain that many visiting scientists came to Oxford for the express purpose of discussing their work with him. Like Elton, he was a natural fuel station. Although exceptionally nice to all, he could be devastatingly frank, in a dry way, when necessary. For example, when he was asked to look over the life-table material for the blue whale, facts obtained from fourteen years of appallingly laborious research, he remarked: "The figures are consistent with any of the following hypotheses: (1) The population has increased, (2) The population has decreased, (3) The population has been stable." This blow was softened somewhat by Elton's comment that when the figures for the six ocean sectors were looked at separately, the one with the highest percentage of young was that within which the blue whale had been protected.

George Leslie was very much a cerebral worker, "playing with the figures" in his head, rather than on paper. John Clarke recalls hearing George utter a deep shuddering groan as he passed him in the corridor of the St. Hugh's huts, where George was standing staring out of one of the high windows, his head in his hands. John rushed into Dennis Chitty's room to ask if George should be given assistance. "Oh don't you worry about George," Dennis told him cheerfully, "it's just his statistics troubling him." Research on complex theoretical aspects of population dynamics carried knowledge of the Bureau's mission into another academic population, as much of Leslie's work was published in *Biometrika*. He made ecologists and mathematicians more aware of the formidable complexities inherent in population ecology, and provided new tools for dealing with them (87–90).

Elton's allocation of one of his few staff posts to a statistician was evidence that he did not, as some ecologists have assumed, hold statistics in contempt. What he did hold in contempt was the spurious impression of exactitude given to poor data by a smoke-screen of algebraic notation. His attitude was pragmatic; there were some kinds of ecological research which could only be advanced by statistical methods, but there were many for which statistical treatment of the figures was unnecessary. The use of limited facilities had to be allocated priorities: "One wishes that some of the elaborate European investigations of dune faunas had given less time to statistical abstractions and cryptical classical terminologies and more to finding out what the animals do and how they do it, in relation to the rather obvious patterns of structure and climate (51). But when Kitty Paviour-Smith began work on the communities within bracket fungi, he knew she would need statistical training and sent her off to do an appropriate course. Later, Kitty reared populations of *Cis bilamellatus* under different conditions of temperature and humidity to determine the effects of these factors on *little r*. With the time and facilities available she could only rear one population under each set of conditions. She plotted the four points and fitted a line by eye, as advised by George. When she showed this to Monte Lloyd, he was horrified: "But you can't do something like that, you must work out the regression properly." Now feeling worried and insecure, Kitty went back to George and told him what Monte had said. "Kitty my dear," George said with a smile, "you mustn't worry about what these young things say!"

Next door, in the EGI, standards of scholarship were just as high, but the working atmosphere was very different, due to the more assertive personality of David Lack, and also, perhaps because of the smaller size of the Institute. There was only one less position on the research side, but there was no proper workshop and no one like D.K. The more restricted scope of the work undertaken also made the place more susceptible to the dominance of its director, even if Lack had not remained rather a schoolmaster at heart. He had no interest or faith in the usefulness of experiments in elucidating ecological problems; after all, the full title of the EGI was the Edward Grey Institute for *Field Ornithology*. It was a happy accident of architectural design that the doorway over which was engraved the motto *Sine experientia nihil sufficienter sciri potest* led into the BAP and not into the EGI.

At afternoon tea there, Lack liked a more formal setting and felt that as leader of the group he should show leadership in conversation. This was fine for the students, with whom Lack was at his most relaxed and charming, but it could be irksome for the senior people like Reg Moreau and H. B. Alexander. Moreau, who was a kindly and good-humored fellow once remarked to John Gibb, "Tell me John. It is *me* or is David really becoming more intolerable every day?" Bill Sladen, who was writing up his classic work on Adelie penguins at the EGI in 1954, told me how the older staff people conspired to pull David's leg by introducing into the teatime conversation subjects they knew more about than he. They brought the subject around to plants one afternoon and Lack had to listen to their learned discussion. He must have done some homework on the topic, for he took the lead next afternoon and continued the discussion. But they "raised the ante" in the game by switching into the Latin names of the plants, again reducing him to being a listener.

I made the mistake of bursting into the EGI tearoom one afternoon without noticing that it was teatime. David was seated at the head of the table pouring tea from an elegant pot into matching china cups and saucers, and he gave me such a nasty look that I hastily retreated. That was soon after the move to the Botanic Garden, and he was still not on speaking terms with me because of an unfortunate encounter in the grounds of St. Hugh's, two years before. I had returned from Wytham in the late afternoon and was carrying trap containers in both hands as I walked to the covered way which led to the two institutes. The setting sun struck me full in the face, and I was temporarily blinded as I heard a soft "Good afternoon," and responded, "Hi there!" There was a scramble of indignant noise in which I heard the words "uncouth" and "can't be civil," and I turned to see Lack making off in a huff. After that he always passed me without recognition, which I regretted, as I respected his vast knowledge of his subject and agreed that I was uncouth.

But after a few more years, when I was visiting Oxford, we met under circumstances which brought us to a fleeting moment of fellowship.

After calling on Arthur Cain at the Pitt-Rivers Museum, I was crossing the vast courtyard in front and ran into David Lack. He was gazing skywards and making agitated body movements. He looked about him and saw me. I was the only person in sight. "Ah. Er. Crowcroft. Look! They are getting ready to leave!" I looked up and saw a swirling cloud of little birds. Although almost completely ignorant about birds, I saw their saber-shaped wings and heard their high-pitched cries and recognized the swifts that Lack and his students had been studying high in the tower of the Museum. It was a fascinating piece of work; they had investigated the composition of the aerial plankton the parents were bringing to the young by stealing some of the regurgitated balls of food before the nestlings could eat them (84). Lack's face was still partially paralyzed from a chilling draft he had endured while watching the birds through a chink. Now the colony was joined by others and formed a swirling nebula of excited little birds. Then, as we watched together, they began to stream away towards Africa. It was a moment full of wonder, and Lack wanted to share it with someone—anyone would do. We parted, glowing with joy, and never saw one another again. Lack was cruelly cut down by illness in his prime.

Charles Elton remained for many years in the relatively junior post of demonstrator. It was a job regarded as a "plum" and had been held by other brilliant Oxford graduates, but they generally moved on after a few years to senior posts elsewhere. When the Department of Zoological Field Studies was created on July 8, 1947, he was given the title of university reader, but he received only a lecturer's pay, as more funds were not in the budget. In January of the following year his post was properly established and he was reader in animal ecology until he retired, at the official retiring age of 67, in September 1967. Within the Bureau he taught, as befitted the abbot of a small enclosed order, mainly by his example and his conversation. But as reader he was obliged, by statute, to do some lecturing to undergraduates. Elton gave two series of lectures: one in animal ecology, and one in zoogeography. The ecology course was at first aimed mainly at forestry students, but it was attended by undergraduates reading zoology, and by a sprinkling of postgraduate people. As Britain's ties with Commonwealth nations became more tenuous, overseas jobs for foresters became scarce, and the teaching of forestry began to wither away. Elton recalls wryly that he had been drawn to forestry himself, but that Professor Poulton had advised Professor Elton that his son might find the subject too difficult, and that he might, therefore, be better advised to read zoology.

There were only eight lectures in the zoogeography course, and from eight to ten in animal ecology. Usually the courses were given in alternate academic years, so it was not a heavy teaching load. At one time the ecology course was expanded to sixteen lectures, but Hardy cut it back to eight when he began to cover much of the material in his own lectures on marine ecology. Elton's lectures were unusually thoughtful and discursive; some found them vague. The first lecture in a series would draw thirty or more. The next week about fifteen would be back and they would stay for the whole series. In presenting facts Elton was less disposed to say "This is the way it is," than "This is the way it appears to be. But it may not really be like that at all!" When he described the distribution of a species on a particular part of a particular plant at a certain time of day in one season of the year, he would demolish the image by gently adding, "That is, of course, if it doesn't happen to rain." Tom Park described Elton's lecture on continental drift as a magnificent discussion of the pros and cons, and one which left him in the position of being able to either embrace the theory or not, with equal enthusiasm. After the lecture, Elton asked him how he felt about continental drift and Park told him. Elton simply said, "I don't know, either."

New jargon was being invented in profusion, especially in the United States (it was also being generated in Germany and Russia but we didn't know about it). Elton refused to use the new terms, feeling that to put a label on something abstract tended to freeze thinking about it. Reviewing a book in *Nature* that abounded in such terms he wrote: "The biochores and superbiochores are finally combined into still higher groupings, which may be called 'biocycles.' Some readers might feel that they need a 'biocycle' to enable them to get around the 'chores' of reading this sort of superbly unnecessary jargon. Phrases like the 'spruce-moose biome' are likely to amuse without illuminating." He regretted indulging his sense of humor in print like this because the author, W. C. Allee, never forgave him. Neither was Andrewartha likely to be forgiving about Elton's assessment of his monumental attempt, with Charles Birch, to assert Australian leadership in ecological theory: "Most of this section about population relationships is written in a highly critical, often polemical, manner, and it forms a sustained and oddly contemptuous attack on what the authors regard as the dogma of the regulation of numbers through varying density-dependent processes based proximately on biotic links. It does not seem likely that the reader will be entirely impressed by the implication that nearly all previous research on the subject is irrelevant to the interpretation of field population changes, or that the study of simplified mathematical models and their testing in the laboratory is misleading." Elton went on to say that the new theory of population regulation put forward by Andrewartha and Birch could be regarded as every bit as theoretical as the theories they blasted: "For all practical purposes at the present time, we know no more about the behaviour in the field of the innate capacity for increase than Malthus did. The notion of innate capacity for increase with its stable age distribution, is really quite as highly theoretical and abstract a tool for study as the equation for the logistic curve, or Volterra's predator-prey theories. Apart from the human material originally used, it is so far based entirely upon laboratory work with populations living in supposedly defined conditions. The authors have, indeed, substituted for what they consider a dogma, another just as theoretical and with rather less common sense behind it. In spite of the air of certainty which pervades this large book, there are striking inconsistencies like this. These may not appear at first reading. The style is lucid and the learning great, but the thoughts behind are not always illuminating."

In his lectures, Elton took great pains not to be dogmatic; the most definite statement I recall was that we must be wary of being definite about anything in animal ecology at this early stage of its development. For example, regarding the use of the term "population control," he said, "There may be different causes of population control of a species, perhaps, in different parts of its range, at different times, or in different habitats." His diffident way of throwing us ideas that were new to us, and which indeed may have been new to anybody but himself, made them easy to miss. "Why do species not become extinct?" he asked, and paused just long enough. "The answer is, of course, that they *do* become extinct—*locally*. Populations fluctuate between over-increase and extinction. In a really small patch, they do become extinct from time to time, but they re-enter the community through movements from elsewhere." In this matter, as in most topics he touched upon, the examples were drawn from species that mankind was seeking to manage in some way, and they were frequently pests. For although concerned with theories in ecology, he constantly returned to the theme that ecology was for practical use in the world. When I arrived at the Bureau, and he asked me what I knew about animal ecology, I answered "Practically nothing, so I'd appreciate your giving me a reading list." "You had better start," he said, "by reading Ed Graham's *Natural Principles of Land Use.*" (I had the good fortune to meet Dr. Graham at a conference later, and he was delighted when I told him of this.)

Although personally engaged in searching for principles underlying the apparent complexity of the natural world, Elton warned us about expecting to be handed down *Laws of Nature*. There are natural laws which enable us to predict what will happen when bodies are in motion and meet, or when chemical reagents are brought together. But when we look at phenomena as dynamic and diffuse as the relationships between species, he warned us: "You cannot guarantee that this or that will happen. But you know that some things will always happen, that some things are more likely to happen than others, and most important, that some things will never happen." This was not a reassuring view of the natural world to give to students, even in Oxford, where they are less likely than students in most other places to expect ready-made answers to predictable examination questions. The same deceptively simple advice has been given to later generations of population ecologists by Richard Lewontin: "It is *not* the function of theory to describe what happened in a particular instance. Only observation can do that. The purpose of theoretical studies in population biology is to *set limits*, to say, 'This can happen; this cannot. This process will be extremely slow or of extremely small magnitude as compared to that process.' Theoretical population biology is the science of the possible; only direct observation can yield a knowledge of the actual . . ." (86).

Elton's lectures were reworked with a degree of industry that no listener would suspect. A day on which he had to give a lecture was a day of purgatory. He shut himself in his room, worked over his notes, and selected the slides to be used from a collection of about a thousand which D. K. had mounted in glass, using the new 35mm format. Then after this conscientious preparation, he delivered the lecture in such a diffident manner that those who did not know him could be forgiven for supposing that he was ill prepared. He sat awkwardly with his head down much of the time, raising it now and then to stretch his neck in a characteristic way, and to peer over his glasses at the rows of examiners and inquisitors in the tiered lecture theater. His voice was naturally soft, and even if the old theater had been equipped with a microphone he would not have used it. The room was absolutely quiet as people froze, straining to hear him. I learned, from attending his lectures more than once, that it was when he dropped his voice and became least audible that he made his most profound and original remarks.

Showing slides was a relief, and his voice strengthened. The slides were intimately related to what he had to say, but the connections could be subtle and idiosyncratic. There were no carousel projectors in those days; the very latest from the J. Arthur Rank Organization had a track for two slides, the one being removed and replaced by hand when the other was in front of the lens. Most Oxford lecturers, in formal suits, instructed their minions, "Next slide please," but for Elton in his old cardigan the slides silently appeared just when he wanted them. John Clarke was intrigued by this efficiency and solved the puzzle by sitting immediately behind D. K., who was always the Boss's projectionist. There was a tiny light bulb mounted in a long narrow tube directed so that only D. K. could see it, and there was a wire running to the Boss's table.

Elton had no false illusions about his ability as a public speaker. He wrote to me on this subject: "I was not a good lecturer, and it was hell preparing and giving them. My mind used to white-out totally the day before; but John Wittaker, discussing the question and agreeing with my poorness added: 'But we always felt *you had been there,*' referring to my personal examples and knowledge of people who had done things. And after all, the Schools' papers never asked more than one or two questions on ecology."

There was a course in field ecology at the Bureau for one week in September almost every year between 1948 and 1956. Elton was encouraged to do this (he says, "prodded into it") by Tom Park, who conducted such courses in the University of Chicago, where they had been pioneered by Victor Shelford. Tom Park spent the better part of 1948 at the Bureau as a Rockefeller Foundation Fellow. He felt that his "ecological perspective" gained much from talking with Elton and the other staff members and through the use of the unique Bureau library. While in Oxford, Park worked increasingly with George Leslie, who became interested in the classic "milk-bottle" *Tribolium* populations developed in Park's laboratory. This led in later years to several joint papers concerned with the experimental and stochastic aspects of interspecies competition (93, 94). While in Oxford in 1948 Park assisted in the formation of the first field course. Almost all of the students taking part in the course were from Imperial College, London University. The college had strong courses in natural history, especially entomology. Tom Park recorded his impressions in his personal diary and extracted them decades later for my "diplomatic use."

AUGUST 19: Talked before the Bureau's Thursday meeting on ideas of teaching graduate ecology—outlining largely the rationale and approach of 304 (my course at U. of C.). I think this went over well.

AUGUST 20: Elton came in to say he enjoyed my talk very much. He claimed the Bureau needed to know this could be done but he didn't think he could put in so much effort. Claimed he is a bad teacher: "I have the knowledge— my staff (meaning Chitty and Southern primarily, I guess) don't have the knowledge but are good teachers."

SEPTEMBER 10: . . . with students coming next Wednesday for a week's instruction in ecology, we are beginning to work at the job. Charles has invited me to participate according to my desires. The students will go to Wytham Woods each morning, and work in the laboratory each afternoon with a conference at tea time. . . . Charles and I went to Wytham Woods where I helped him plan his first class exercise on forest-floor quadrats in two sorts of leaf-litter. I then "sold" him on a log-succession study—we return to WW tomorrow to plan this. Charles seems very glad to get pedagogical suggestions. Despite his intimate experience in field work, he has not had much experience in field teaching. He has great knowledge, however, about animals and plants; what they are doing and where they are doing it.

SEPTEMBER 13: Gave Charles an outline of suggestions for his log-succession study. . . . Also attended afternoon tea at which event the 16 students here to take the ecology class were present. Charles made several generous comments to them about my help in planning the course.

SEPTEMBER 15: Spent most of the day with the class in Wytham Woods. We studied grasshopper distribution relative to habitats and applied the marking and release "Lincoln Index" to a census of grasshopper populations in the afternoon. . . . On the field trips, in addition to the students, were Elton, Southern, Freeman, Hobby (Hope Department), and myself (GEORGE wouldn't go near the place!) The students are quieter than ours. It is hard as yet to judge their relative capabilities. Elton conducts the course in a graceful, informal way with a minimum of direction. I rarely heard a laugh or a joke from the students. This may only mean that they feel strange, or, it may illustrate so-called English reserve. After the trip Dennis Chitty talked in the library about the Lincoln Index—he talked well too.

SEPTEMBER 16: Spent the morning with the class in WW. I had two pairs of students placed in my care. They worked well and were quiet and polite. I don't think they have the background of my group at U. of C. but they are as intelligent.

SEPTEMBER 17: Morning with the class—this time studying the forest-floor litter. I took charge of half of the class. Professor Hardy called on me this afternoon—he expressed interest in ecological teaching, both here and at Chicago. . . . In the evening I joined the class on a two hour trip to WW, 8 to 10 PM. We staked off quadrats and watched and counted active forms on the surface of the litter. The students enjoyed this experience in their quiet way. (I missed the campfire and the steaks, however).

SEPTEMBER 18: . . . Went out in the morning with the class to WW. They studied the dispersion of several species in the herbs (dog's mercury) by sweeping x number of times in y places in z number of intervals.

SEPTEMBER 20: . . . The trip was under the supervision of Peter Hartley, who lectured to the students in his best military manner (loud and clear). He combined enthusiasm, preaching, good humor, and lucidity. The field study was designed to explore the nesting and feeding heights of four species of tits (birds!).

SEPTEMBER 21: Last field trip of the course. Studied the fauna in the leaf-litter of a tiny brook. (I had urged Charles to get a bit of water in the course.) Class ended at tea with a brief summary of the points that we had covered.

SEPTEMBER 23: . . . Spent a good part of the afternoon writing a critique of the ecology course for Elton, which he seemed glad to get. He said he agreed with most of my comments.

The field ecology course was held for the second time in 1949 with only eight students, and similar numbers were invited to the classes in the following years. Richard Freeman, who had been a wartime staff member and was now lecturing at University College, acted as Elton's stand-in for most of the exercises, and Mick Southern took over the grasshopper census and Lincoln Index project on Rough Common. The field ecology course was held only seven times and sixty-two students had the benefit of the experience (Appendix 3). By the time the course

The 1953 field ecology course in Wytham. From left, Charles Elton, G. Blane, Mick Southern, Amyan Macfadyen, Maxim Todorovic, D. Duckhouse, M. Boyd, T. Myers, T. Bagenall, J. Lock, and Richard Freeman. Photograph by Denys Kempson. Courtesy of Charles Elton.

was dropped, a number of BAP "graduates" were teaching similar courses in various parts of the world, and animal ecology was well established in the curricula of a number of British universities, especially London, Aberdeen, and Durham. This spread of ecological teaching was not entirely due to the Bureau's efforts, of course. Always wary of spreading his resources too thinly, Elton now felt that he could give up the annual commitment by himself and the staff. In retrospect it is obvious that if he had kept it going, perhaps turning the whole thing over to Southern and Freeman, the Bureau would not have been so vulnerable to Pringle's predation at the end of the following decade.

In the 1949 course I had the good fortune to be paired with a fine young naturalist, Owen Gilbert, from Manchester University. He had a good working knowledge of the invertebrates we extracted from our samples and our log (they were all strangers to me although I could place many in their right Order), and also an interest in theoretical ecology (64). After getting a Ph.D. a few years later, Owen decided, unfortunately for the subject, that ecology was a great hobby but not sufficiently serious to devote one's life to. After flirting with medicine, he went into the steel industry and applied the methodology he had learned with animals to the study of activity rhythms and daily movements of steel production workers.

The secondary function of the Bureau—to provide a clearing house for information about animal populations—involved keeping open house in the library, as well as storing material there. The comings and goings of visitors were not announced in advance, however eminent they might be in their own territories. We usually found them at the bare plank tea-table when we got back from Wytham, so we just dropped into a vacant chair and introduced ourselves. Elton might appear with someone who interested him and make introductions to those immediately adjacent, but there were no formal introductions or presentations and explanations. I recall coming back from trapping in my field gear and finding myself seated next to a ruddy-faced gentleman in a dark formal suit. "Hallo, my name's Crowcroft" I announced cheerfully, "what's yours?" He looked a bit startled and replied rather stiffly, "Er, Nicholson." I sat up and turned to him with enthusiasm. In Elton's lectures I had heard about some classical work on blowfly populations by A. J. Nicholson of CSIR. "Not *the* Nicholson?" I asked. He looked both pleased and embarrassed and relaxed a little, "H'm, 'hem, well, A. J."

This informality did not suit all of our guests. Simon Bodenheimer had been given handsome acknowledgment in *Voles, Mice and Lemmings* for his pioneer work on voles in Palestine. In spite of his fervent Zionism and pride in being three times wounded in "the war of independence," he remained a German professor until the end of his days. So he expected red-carpet treatment and never got over the informality he encountered at the Bureau. When he reissued his classic book *Problems of Animal Ecology* as *Animal Ecology Today,* he dedicated it to "Four great pioneers of Animal Ecology, Victor Shelford, Richard Hesse, A. J. Nicholson, and Paul Errington." He made a point of opening the copy he gave me and reading the dedication. Then we chatted about some small-mammal matters and I brought the subject around to the Bureau. I asked him what he thought would happen when Elton retired or died. Bodenheimer assumed a pose of exaggerated astonishment, "But my dear Dr. Crowfort, did you not know? Elton has been dead for yearsss."

Visitors, whether casual or long-term, were not pressured into talking about their work, but some felt obliged to offer and a few felt moved to insist, sometimes when their knowledge of English was as inadequate as our knowledge of their own native tongue. There was no rigid routine

Tom Park presents Charles Elton with an Eminent Ecologist award. This was not the actual certificate from the Ecological Society, but a cartoon by Ken Marsland. The occasion was marked by having cakes for tea instead of the usual plain cookies. Photograph by Denys Kempson. Courtesy of Charles Elton.

of in-house seminars, but each year there were ten or twelve, mostly in spring or fall, usually by Bureau inhabitants as an aid to getting their thoughts in order and to let their colleagues and mentors know what they were doing. If speakers did not seem to know their subject, or got out of depth by trying to appear too knowledgeable, they were liable to be cut up in the polite and relentless Oxford style.

It was in this low-key setting of the Bureau library at teatime that Elton became a superbly stimulating teacher, and the impressions that were carried away to enhance the Bureau's reputation as a hothouse for ideas were mostly created at such times. The eager visitor who buttonholed the Boss at other times and subjected him to intensive interrogation or exposition was likely to be cut short. Several visitors have asked me what went wrong: "We were getting on fine, and suddenly he just turned and walked away. I must have said something to offend him." This sudden switching off had happened to me on several occasions. Once when he was trapped with me in a railway carriage, and I was prattling away trying to impress him, he suddenly snapped his head away and stared out of the window until we reached Paddington Station and he could make his escape. So I had come to understand how he felt and was able to reassure people that they had not given offense and that if they had, it was quite ephemeral: "He just couldn't stand what you were doing to him any longer, and had to get away. After all, he is a genius, and he's the nicest one of those you are likely to meet!"

Elton kept a log with brief entries of visitors and their fields of interest, and gave me a copy of those years in which my own involvement with the Bureau was greatest (Appendix 4). He was reluctant to hand over any more, which was a pity because just about everyone who was anybody in ecology made a pilgrimage to the Bureau of Animal Population at some time in their career. The sample shows the international nature of the visiting population. The reputation and influence of the Bureau were spread by the mature scholars who came to expose their ideas to Elton or his senior staff and to use the unique library. They were the people who sent their best students to share their experience and to marvel at the minute nature of the Bureau's scanty physical resources.

Interaction with this flow of top people from different countries was the most valuable component of the Bureau experience. We were like the members of a small tribe inhabiting a remote oasis on the route of many caravans; when the travelers came to refresh themselves at our spring they repaid us with tales of their ecological adventures in distant lands. Many of the acquaintanceships we made ripened into lifelong friendships, so enriching our lives that we came to feel fortunate indeed to have become Elton's ecologists.

Appendix 1: BAP Staff Members, Other Research Workers, and Resident Visitors

NOTE: * indicates member of staff; † = died.

*Elton, C. S.	1932–67	Oxford University. Retired 1967
*Middleton, A. D.	1932–37	Stowe School, Bucks.
	1939–45	Seconded from ICI. †1986
*Davis, D. H. S.	1932–35	Oxford University. †1986
*Ranson, R. M.	1932–44	Stowe School, Bucks. †1944
*Warwick T.	1932–35	Manchester University
*Leslie, P. H.	1935–67	Oxford University. †1972
*Chitty, D. H.	1935–61	Toronto University
Keay, Miss G.	1935–36	Oxford University. †1967
Swynnerton, G.	1935–36	Oxford University. †1959
*Gibbs, Miss M.	1935–67	Oxford
*Chitty, Helen	1936–61	Toronto University. †1987
*Wells, A. Q.	1936–39	Oxford University. †1956
Evans,F. C.	1936–39	Haverford College Pa.
	1962–63	Michigan State University
*Southern, H. N.	1938–67	Oxford University. †1986
*Hewer, H. R.	1939–41	Imperial College. †1974
*Venables, L. S. V.	1939–45	Wartime volunteer †1989
*Freeman, R. B.	1939–46	Oxford University. †1986
Wykes, Miss U. M.	1939–42	Wartime volunteer
Fisher, J. M.	1940–44	Oxford University. †1970
*Nicholson, Mrs. Mary	1940–43	London
*Rzoska, J.	1940–46	Poznan University. †1984
*Vevers, H. G.	1940	Oxford University. †1988
*Watson, J. S.	1941–49	Oxford University
	1955	Department of Scientific and Industrial Research, New Zealand. †1959
*Thompson, H. V.	1942–46	University College, London
*Perry, J. S.	1943–46	University College of North Wales
*Laurie, Miss E. M. O.	1943–48	University College of North Wales
*Shorten, Miss M. R.	1943–48	Oxford University
*Kempson, D. K.	1945–67	Oxford. †1977
Larkin, P. A.	1946–48	University of Saskatchewan
	1961–62	University of British Columbia
Macfadyen, A.	1947–49	Oxford University
* "	1949–56	University of Coleraine
Birch, L. C.	1948 (part)	University of Sydney
Ratcliffe, F. N.	1948 (part)	Oxford University (CSIR) Council for Scientific and Industrial Research

Park, T.	1948–49	University of Chicago
Crowcroft, W. P.	1949–50	University of Tasmania
Clarke, J. R.	1949–52	Oxford University
Godfrey, Miss G. K.	1949–51	Oxford University
Miller, R. S.	1949–51	University of Colorado
* "	1951–61	
* "	1963–64	Yale University
*Macfadyen, Usrula	1949–51	Oxford University. †1986
Overgaard-Nielsen, C.	1950–52	University of Copenhagen †1989
Phillips, Miss W.	1950–51	University College of North Wales. †1980
Duffey, E.	1950–52	Leicester University
Main, A. R.	1951 (part)	University of Western Australia
*Macpherson, Miss B.	1951–55	University College, London
Williams, G. R.	1951–53	Department of Scientific and Industrial Research, New Zealand
*Bowen, Miss U. H.	1952–53	Oxford University
Brereton, J. LeGay	1952–55	University of Sydney
Davies, M. J.	1952–55	London University
Jenkins, D.	1952–56	Cambridge University
Tener, J. S.	1952–53	Canadian Wildlife Service
Todorovic, M.	1952–54	University of Belgrade
Fager, E. W.	1953–55	University of Chicago
	1967 (part)	Scripps Institute of Oceanography. †1976
Falls, J. B.	1953–54	University of Toronto
Huxley, T.	1953–56	Oxford University
*Dobbs, Miss B. T.	1954 (part)	Oxford University
Lowe, V. P. W.	1954–57	Oxford University
	1966–67	Institute of Terrestrial Ecology
Southwick, C. H.	1954–55	University of Wisconsin
Paviour-Smith, K.	1955–58	Otago University, New Zealand
* "	1958–67	
Eccles, D. H.	1955–59	University of Cape Town
Kikkawa, J.	1955–57	Kyoto University
Efford, I. E.	1957–60	Leicester University
"	1961–62	Environment Canada
Dawson, Miss C. M. J.	1957–60	Oxford University
Newson, R. M.	1957–60	Cambridge University
Pitelka,F.	1957–58	University of California, Berkeley
Lloyd, M.	1957–59	University of Chicago
* "	1959–62	
Odum, E. P.	1958 (part)	University of Georgia
Batchelor, C. L.	1958 (part)	New Zealand Forestry Service
Gall, H. J. F.	1958–59	University of Chicago
*Hurry, S. W.	1960 (part)	Cambridge University
Murdoch, W. W.	1960–63	Glasgow University
Krebs, C. J.	1960–61	University of British Columbia
Smyth, M. E. B.	1960–63	Adelaide University. †1974
Ghelardi, R. J.	1960–62	Scripps Institute of Oceanography
Pimental, D.	1961 (part)	Cornell University

*Elbourn, C. A. D.	1961–67	Sir John Cass College, London
Trapido, H.	1962–64	Rockefeller Foundation
Sutton, S. L.	1963–66	Oxford University
Watts, C. H. S.	1963–66	Adelaide University
Mathews, C. P.	1963–66	Swarthmore College, Pennsylvania
*Whittaker, C. P.	1963–66	Durham University
Buckner, C. H.	1965–66	Canadian Forest Service
Peterson, R. S.	1965–66	Johns Hopkins University. †1969
Schindler, D. W.	1965–66	University of North Dakota
Love, Miss D. M.	1965–67	Oxford University
Schindler, J. E.	1966–67	University of North Dakota
Kitching, R. L.	1966–67	Imperial College, London

Appendix 2: **BAP Doctoral Theses Deposited in the Elton Library**

Evans, F. C. 1939. Habitat selection in small mammals, with special reference to Rodentia and Insectivora. CN 31/3/1

Perry, J. S. 1945. The reproduction of the wild Brown rat. CN 31/3/1*

Larkin, P. A. 1948. Ecology of mole (*Talpa europea* L.) populations. CN 31/6/1

Chitty, D. H. 1949. Factors controlling density of wild populations, with special reference to fluctuations in the vole (*Microtus*) and the Snowshoe rabbit (*Lepus americanus*). CN 31/5/1–2

Miller, R. S. 1951. Activity patterns in small mammals, with special reference to their use of natural resources. CN 31/7/1–2

Clarke, J. R. 1953. The response and behaviour of animals at different population densities, with special reference to the vole (*Microtus agrestis*). CN 31/9/1–2

Godfrey, G. K. 1953. Factors affecting the survival, movements, and intraspecific relations during early life in populations of small mammals with particular reference to the vole. CN 31/8/1

Crowcroft, P. 1954. An ecological study of British shrews. CN 31/10/1

Davies, M. 1955. The ecology of small predatory beetles, with special reference to their competitive relations. CN 31/11/1

Fager, E. W. 1955. A study of invertebrate populations in decaying wood. CN 31/12/1

Brereton, J. 1955. A study of factors controlling the population of some terrestrial isopods. CN 31/13/1

Duffey, E. A. G. 1955. An ecological study of the spider (Araneae) communities in limestone grassland CN 31/14/1**

Jenkins, D. 1956. Factors controlling population density in the partridge. CN 31/15/1

Paviour-Smith, K. 1959. The ecology of the fauna associated with macrofungi growing on dead or decaying trees. CN 31/17/1

Efford, I. E. 1960. A population study of water-mites (*Hydracarina*). CN 31/18/1

Newson, J. 1960. The relationship between behaviour, population density and physiological condition in voles (*Microtus agrestis* and *Clethrionomys glareolus*). CN 31/19/1

Newson, R. 1960. The ecology of vole and mouse populations in different habitats. CN 31/20/1

Smyth, M. 1963. The effects of varying abundance on the population dynamics of rodents, with special reference to the Bank vole, *Clethrionomys glareolus*. CN 31/21/1

Murdoch, W. 1963. The population ecology of certain carabid beetles living in marshes and near fresh water. CN 31/22/1

*Ph.D. University College of North Wales
**Ph.D. University of London

Schindler, D. 1966. Energy relations at three trophic levels in an aquatic food chain. CN 31/25/1

Sutton, S. L. 1966. The ecology of isopod populations in grassland. CN 31/26/1

Watts, C. H. 1966. The ecology of woodland voles and mice with special reference to movement and population structure. CN 31/27/1

Mathews, C. P. 1968. The exploitation of food resources by *Gammarus pulex* L. and other animals in woodland streams. CN 31/56/1

Kitching, R. L. 1969. The fauna of tree-holes in relation to environmental factors. CN 31/33/1

Schindler, J. E. 1969. The food and feeding ecology of zooplankton populations in a small reservoir. CN 31/32/1

Appendix 3: Students Attending BAP Field Courses, 1948–1956

SEPTEMBER 1952

J. F. Barker	Sheffield University
M. J. Delany	University College, Exeter.
J. H. Connell	University of California; Glasgow
W. J. Burley	University of Oxford
P. W. Carden	University College of North Wales, Bangor.
J. C. Coulson	University of Durham
Miss T. Dobbs	University of Oxford (BAP)
Miss M. Bennett	Infestation Control Division, Ministry of Agriculture, Fisheries and Food

SEPTEMBER 1953

M. Todorovic	University of Belgrade (BAP)
G. Blane	Southampton University
T. Myres	Cambridge University (EGI)
J. Morton Boyd	Glasgow University
A. F. G. Dixon	University College, London
D. Duckhouse	University College, Leicester
J. Lock	Bristol University
T. Bagenal	Marine Station, Millport

SEPTEMBER 1954

V. P. W. Lowe	Oxford University (BAP)
Miss S. Middleton	Southampton University
P. D. Gabbutt	University College, Exeter
J. E. Peachey	University of Durham
W. G. Potter	Oxford University
G. Surtees	Durham University
R. G. Allan	Edinburgh University
A. J. Pontin	Oxford University (Hope Dept.)
J. LeG. Brereton	BAP (part of course)

No field course
was held in 1955.

SEPTEMBER 1956

Miss K. Paviour-Smith	Otago University (BAP)
Miss J. Woodward	St. Andrews University
J. Kershaw	Manchester University; School of Agriculture, Cambridge
J. Kikkawa	Kyoto University (BAP)
J. M. Cherrett	Durham University
B. Heighton	Durham University
J. Pollack	University College, London
I. E. Efford	University College, Leicester (BAP)

Appendix 4: Department of Zoological Field Studies

Bureau of Animal Population Diary of Events for 1949–1952, by C. S. Elton

1949

Jan 26
: Mr. Oliver Hook came to obtain information about how to conduct a mammal survey of the New Forest. He is a retired engineer, living at Brockenhurst, who has started a very energetic programme of general survey.

Feb 22–Mar 12
: Miss Winifred Phillips, who is working on rabbit populations in Wales, under a grant from UFAW, worked with Southern in Wytham Woods, and lived at the Chalet, also using the library.

Mar–Apr
: W. Potter, Magdalen College School, stayed at the Chalet, and helped Southern with breeding census on owls.

Mar
: Peter Crowcroft, arrived from the University of Tasmania to do training and research, and started as field assistant to Southern.

Mar 28
: Dr. Gerard R. Pomerat, Assistant Director for the Natural Sciences, European Sector, under the Rockefeller Foundation, paid an official visit to find out what the Bureau was like.

Apr 1–2
: Professor Thomas Park, Department of Zoology, University of Chicago, called at Oxford on a return visit from America, working on behalf of the State Department, on a report upon British ecological activities.

Apr 4–9
: Mr. S. A. Barnett, Infestation Division, Ministry of Agriculture and Fisheries, was working in the library on rat control problems.

Apr 4–13
: Mr. R. B. Freeman, Department of Zoology, University College, London, was working in the library, completing bibliographical work on the wartime research on rodent control.

Apr 23–May 5
: Dr. L. E. Wagge, lecturer at the Imperial College of Science, London, and two students, Priest and Lamerton, helped with the trapping campaign in Wytham (the two students lived in the Chalet). Dr. Wagge spent some time on the literature in the library.

June
: Dr. Robert Glen, Department of Agriculture, Ottawa, and Dr. A. T. Arnison, Department of Agriculture, Saskatoon, called to consult Elton about the planning of an ecological survey of animals (mainly insects) in the Middle Western Canadian Range lands.

June 20–July 28
: Mr. R. C. Muir, Birmingham University student came to spend a month working in Wytham Woods, living at the Chalet, helping Southern.

June 22
: Professor Bush, Department of Zoology, University of Natal, Petermaritzburg, called.

July 25   Dr. F. W. Preston, Butler, Pennsylvania, made a visit to discuss population trends in man. He is a consultant on the physics of glass manufacture who runs a research and teaching lab of his own on bird ecology.

Aug   W. Potter, Magdalen College School, stayed at the Chalet, helped Southern with the August trapping.

Oct   The following graduates started research work at the Bureau: John R. Clarke (St. John's), zoology graduate Oxford, 1949. Rhodes scholar, University of Western Australia, Perth. (Ecology of the vole).

Miss G. K. Godfrey (St. Anne's), Oxford zoology graduate, 1949. (Biology of the early stages of the vole).

Richard S. Miller (New College), University of Colorado, zoology graduate 1949, under Professor Pennak. Fulbright scholar.

Oct 17–29   A. P. Buxton came to spend a week or two learning the Wytham trapping technique and reading, before going to East Africa to study ecology of monkeys that carry yellow fever.

Oct 26–27   Dr. G. Evelyn Hutchinson, Yale University, called and spent some time in general discussions on physical ecology and the relation of human populations to resources.

Nov. 5–15   Dr. Ellinor Bro Larsen, lecturer in terrestrial ecology at the University of Copenhagen, came for 10 days in order to brief herself on some aspects of dynamic ecology. Her main research is in community ecological survey, and the tolerance limits of insects.

Nov. 7   Mrs. Ursula Macfadyen, began part-time work, assisting in the arrangement of ecological museum collections.

Nov. 17   Lt.-Colonel W. B. L. Manley, Nature Conservancy, went to Wychwood Forest with Elton, and saw the Bureau's library and indexes.

Nov. 21   Mr. Hugh Gordon, of the McMaster Animal Health Laboratory (CSIR), Sydney, Australia, called to see the Bureau and discussed his long-term research on populations of nematode work parasites of sheep.

Nov. 21–Dec. 2   S. A. Barnett of the Ministry of Agriculture & Fisheries, Infestation Division, spent eleven days consulting the Bureau's records, etc.

Nov 25–   Professor Sewall Wright, University of Chicago, visited.

1950

Jan 19   Dr. C. Overgaard Nielson, University of Copenhagen, came for two years.

Jan 31   Miss W. M. Phillips, Aberystwyth, under grant from the Animal Health Trust, until the end of September. (Revision of previous field reserach on rabbits.)

Feb 8–11   H. V. Thompson, Ministry of Agriculture and Fisheries, Infestation Division, came to consult the Bureau's records.

Feb 21–22   Miss Margaret Lawrence, Mr. Worrall, Mr. Chudley, of the Ministry of Agriculture & Fisheries, Infestation, came to discuss methods of estimating mouse infestation in bulk grain stores.

Mar 1–May 26   Dr. E. W. Bentley, Ministry of Agriculture & Fisheries, Infestation Division, came officially to learn about rodent ecology.

Mar 16–May 6   Miss Marie Stephens, an agricultural research graduate from Bangor, Aberystwyth, working on rabbit populations under the Animal Health Trust.

Apr 3   Mr. George Arthur, leading Orcadian naturalist, from Kirkwall, main Island of Orkney. He collected the original Orkney vole stock that Mr. A. Wood sent to the Bureau in 1935.

Apr 2–9   Dr. A. J. Haddow from the Colonial Medical Research Service, Kenya, who is studying yellow-fever-mosquito-monkey relationships in tropical forests, came to learn small-mammal trapping and marking technique and use the library.

May 14–16   Dr. Marston Bates, the Rockefeller Foundation, New York, visited for two days to discuss human ecology problems.

May 15–21   S. A. Barnett of the Ministry of Agriculture & Fisheries, Infestation Division, spent 7 days consulting the Bureau's records, etc.

May 22–June 10   Miss L. M. Kellas, Tsetse Research Organization, Tanganyika, working up her field results in the ecology of the dik-dik.

June 30   Visit from some delegates of the Commonwealth Review Conference, including:
Dr. Siemons (Entomologist from the Department of Agriculture, Canada); Dr. Marias (Principal, Stellenbosch University); Professor Kirkpatrick, (Entomology, Trinidad); Dr. Anderson (soil deficiency botanist, Australia. Formerly Waite Institute); Mr. McKerrich and Mr. Kinsey (Agricultural Chemist and Administrator, Southern Rhodesia); Mr. Callahan (Head, New Zealand DSIR); Mr. Hill (Director, Bureau of Pastures and Field Crops, Aberystwyth); Mr. Ford Robertson (Commonwealth Bureau of Forestry, Oxford), and his assistant, Mr. Beak; Mr. J. D. Peak (Assistant Secretary, London, H. Q. Commonwealth Agricultural Bureau).

June 15–24   Miss N. Lonsdale, Lady Margaret Hall, and Miss J. Hennessey, Lady Margaret Hall, assisting Mr. Southern.

July 4   Mr. F. Wilson, Division of Economic Entomology, CSIR Australia.

July 10   Dr. T. Weiss-Fogh, Professor A. Krogh's Laboratory, Copenhagen.

July 17   Professor F. S. Bodenheimer, Department of Zoology, The Hebrew University, Jerusalem.

July 24   Dr. R. Spärck, Director, Zoological Museum, Copenhagen, Denmark.

July 25   Dr. Karl P. Schmidt, Chief Curator of Zoology, Chicago Natural History Museum, Lecturer in the Department of Zoology, University of Chicago.

July 27–Aug 5   Dr. N. Haarløev, Zoologisk Laboratorium, Landbohöjskolen, Copenhagen.

Sept 26   Mr. Eric Duffey, a graduate of Leicester University, with some years' Fleet Air Arm service and experience in ornithology (including an expedition to Bear Island) came under a Nature Conservancy grant to train, and is going to do research on the spider populations of woodland etc.

Oct 5	Mr. N. H. Taylor, Soil Bureau, Wellington, New Zealand, called to discuss methods of catching rabbits in an area where they don't use burrows. He talked to H. N. Southern and Miss Phillips.
Oct 10–28	Mr. Kaj Westerskow, Vildbiologisk Station, Kalø Pr. Rønde, Denmark, spent two weeks to use the library index on partridges etc. He has done an ecological study of the black grouse in Denmark, and is now doing the common partridge.
Oct 27	Dr. Lionel A Walford, Head of the Fisheries Biology Branch (Research) of the US Fish and Wildlife Service called.
Nov 6–11	Dr. E. W. Bentley, Ministry of Agriculture & Fisheries, Infestation Control Division, visited to use the library.
Nov 17	Dr. Miles, Department of Pathology, Cambridge, called to see H. N. Southern, to consult him about virus diseases. He is going to Australia to work on small mammals as reservoirs of viruses.
Nov 29	Dr. Strickland, Rothamstead Experimental Station, Harpenden, called and saw A. Macfadyen and Dr. C. O. Nielsen.
Nov 30	Mr. G. R. Fenton, Nature Conservancy, called and saw A. Macfadyen and Dr. C. O. Nielsen, and discussed methods of extraction of soil animals and culture techniques.
	Miss Gillian Evans (Mrs. Matthews) Department of Zoology, University College, Leicester, called to discuss population research with reference to her work on intertidal molluscs.
1951	
Feb 8–9	Mr. W. S. Richards, Applied Entomologist, Sudan Government, visited the Bureau and had discussions with A. Macfadyen.
Feb 26–Mar 24	Mr. F. N. Wright, entomologist under the Colonial Office going to Gold Coast, spent a month reading and picking up methods.
Mar 5–10	Dr. Vernon Link, US Public Health Service, in charge of control survey and research on sylvatic plague in the US, stayed and also gave a talk on plague.
Mar 7	Dr. N. A. Mackintosh, Director of the Discovery Investigations called to discuss age distribution figures for catchers of the Southern blue whale.
Apr 23	Dr. D. C. Swan, Waite Institute, Adelaide, South Australia, visited and gave some account of his war-time scrub typhus work on a small island off New Guinea.
May 7	Mr. Victor Cahalane, Assitant Chief, United States National Parks Service, on his way to America after a long tour of African Parks and Game Reserves.
May 8	Mr. Pickering, Reckitts Bros., Hull, called and used the library.
May 22	Mr. Carl Edelstam, Department of Zoology, University of Stockholm called.
June 6	Mr. Ollerenshaw, starting research for the Veterinary Laboratory, Ministry of Agriculture and Fisheries, Weybridge, Surrey, for a consultation about the ecology of the sheep liverfluke snails.
June 25–Jly 14	Miss N. Lonsdale and Miss J. Hennessey, students from the

	Department of Zoology, Oxford, assisted Southern with his owl pellet work.
Jly 1–31	Mr. Richard Vaughan, lived at the Chalet and assisted Southern in his owl work.
Jly 2–6	Mr. J. L. Harrison from the scrub typhus research team at Kuala Lumpur, Malaya, visited and used the library.
Jly 16–end of Sept.	Mr. A. R. Main, of the University of Western Australia, Perth, came to spend two months in the Bureau, after a year's study in the Zoology Department, University of Chicago.
Jly 17	Mr. G. Saunders, a leading farmer in southern New Zealand, called to ask about the rabbit problem.
Jly 13–14	Mr. S. A. Barnett, Department of Zoology, Glasgow, visited to use the library.
Jly 30	Professor George Clarke, Harvard University, Cambridge, Mass. temporarily working in Oslo, to discuss a book on ecology he is writing. He also has a part-time post at Woods Hole, Mass.
Jly 31	Dr. J. J. Gryse, Canadian Forest Insect Survey; Dr. C. W. Farsted, Field Crop Insect Laboratory. Lethbridge, Alberta, head of large Dominion laboratory; Dr. Robert Glen, Department of Agriculture, Ottawa.
Aug 1	Dr. A. M. Davies, Micro-biological Laboratories, Weizmann Institute of Science, Rehorot, Israel, to discuss the setting up of epidemiological experiments on voles or mice.
Aug 1	Miss Brenda M. Macpherson started a year's work indexing the Wytham Ecological Survey records under a Nature Conservancy grant. R. S. Miller is staying on in charge of this work for 10 months.
Aug 15	Mr. G. N. Sale, Forestry Officer of the Nature Conservancy. Reconnaissance of woods in the Dell Coppice area, with C. Elton and R. S. Miller.
Aug 16	Mr J. E. Satchell, Rothamsted Experimental Station, Harpenden, Herts.
Sep 24	Dr. C. R. Twinn, Head of the Household & Medical Entomology Unit, Canadian Department of Agriculture, and Mr. J. L. Hichon, Fungicide & Insecticide Research Co-ordination Service.
Oct 11–15	Mr. H. Roberts, who is working at the Department of Zoology, University College, Southampton, under a Nature Conservancy Maintenance grant, spent several days reading and discussing his project for a community survey of a beechwood in the New Forest.
Oct 19	Dr. L. Martin, Institut Agronomique de Gembloux, Belgium.
Oct 29–30	Mr. R. A. Davis and Mr. R. J. Clarke, Ministry of Agriculture and Fisheries, Infestation Control Division.
Nov 1–3	Mr. R. Rose-Innes, Plague Research Laboratory, Department of Public Health, PO Box 1038, Johannesburg, Union of S. Africa.
Nov 5–6	Mr. Allen Brooks, Department of Zoology, University of Toronto, Canada.

Nov 5	Mr. Gordon R. Williams, Wildlife Service, Department of Internal Affairs, Wellington, New Zealand.
1952	
Jan 1	Moved from 91 Banbury Road to Botanic Garden, High Street.
Feb 8	Mr. Peter Scott, The Severn Wildfowl Trust, visited the Bureau and discussed analysis of marking and recapture data for geese.
Feb 13	The Registrar, Mr. Veale came to tea.
Feb 20	Nine members of the Edward Grey Institute came to tea.
Mar 5–13	Mr. J. Andersen, Vildtbiologisk Station, Kalo, pr. Rønde, Denmark, visited to use the library and have discussions.
Mar 19	Dr. R. Newton Clark, World Health Organization, Geneva.
Mar 28	H. V. Thompson, Ministry of Agriculture & Fisheries, Pest Infestation Control Division, to use the library. Dr. C. Overgaard Nielsen left the Bureau to take up work as Director of Mols Laboratoriet, Femmøller, Denmark.
Apr 1–6	Dr. Paul Pirlot, Institute pour la Récherche Scientifique en Afrique Centrale, Costermansville, Belgian Congo, working on small-mammal country surveys in tropical forest, came to learn methods and use the library. He saw Wytham trapping routine.
Apr 5–26	Dr. W. H. R. Lumsden, Virus Research Institute, PO Box 49, Entebbe, Uganda—as above.
Apr 27	Mr. P. W. Crowcroft, Ministry of Agriculture & Fisheries, Pest Infestation Control Division, for six weeks.
May 5	Mr. L. R. Clark, Division of Entomology, CSIR, Canberra.
May 6–8	H. V. Thompson, Ministry of Agriculture & Fisheries, Pest Infestation Division, to use the library.
June 13	Dr. & Mrs. Bakken, Wisconsin Department of Wildlife Management.
June 18	Dr. Vernon Joyce, Government Research Farm, Wad Meddani, Sudan.
Jly 1–31	Mr. L. R. Clark, Division of Entomology, CSIR, Canberra.
Jly 1	Dr. Bertha Lutz, an authority on amphibia in the National Museum, Rio de Janeiro, called and was shown Wytham Woods. She watched the ground carefully for snakes and normally carried an anti-venom outfit while doing field work.
Jly 22	Dr. Olof Ryberg, Ålnarp Institute, Akarp, Sweden, a teacher of agricultural zoology, and formerly Chief Inspector of Plant Pests in Sweden, spent a morning arranging library exchanges and took specimens of the Bureau's punch cards.
Sep 1	Mrs. Ursula Bowen joined staff as indexer and abstracter for the Ecological Survey.
Sep 2–11	Eight students attended the Autumn Ecology Course.
Sep 15	Dr. W. Kühnelt, Zoologischen Institut der Universität Graz, Austria, a well-known Austrian soil and woodland ecologist.
Sep 24	Dr. Charles E. Palm, Department of Entomology, Cornell University, called and discussed the ecological effects of insecticides and pest population complexes.

Oct 11   The following graduate research students joined the Bureau:
                M. J. Davies; John S. Tener; L. LeGay Brereton; G. M. Dun-
                net; David Jenkins.
Nov 8   Maxim Todorovic, graduate from Yugoslavia, arrived.
Nov 24   Prof. S. Stankovic, Institut de Zoology, Universite de Bel-
                grade, visited and gave a lecture (in French!) on the relict
                fauna of Lake Ochrid.

References

1. Andrewartha, H. G., and L. C. Birch. 1984. *The ecological web*. The University of Chicago Press. 506 pp.
2. Baker, J. R. 1930. The breeding season in British wild mice. *Proc. zool. Soc. Lond.* 1930: 113–26.
3. Baker, J. R., and R. M. Ranson. 1932. Factors affecting the breeding of the field mouse (*Microtus agrestis*). I. Light. *Proc. roy. Soc.* B, 113: 313–22.
4. Baker, J. R., and R. M. Ranson. 1932. Factors affecting the breeding of the field mouse (*Microtus agrestis*). II. Temperature and Food. *Proc. roy. Soc.* B, 112: 39–46.
5. Baker, J. R., and R. M. Ranson. 1933. Factors affecting the breeding of the field mouse (*Microtus agrestis*). III. Locality. *Proc. roy. Soc.* B, 113: 486–95.
6. Brereton, J. LeGay. 1957. The distribution of woodland isopods. *Oikos (Acta. Oec. Scand.)* 8 (2): 85–106.
7. Browning, T. O. 1963. *Animal populations*. London: Hutchinson. 127 pp.
8. Buckner, C. H. 1969. The common shrew (*Sorex araneus*) as a predator of the winter moth (*Operophtera brumata*) near Oxford, England. *Can. Ent.* 101: 370–75.
9. Carr-Saunders, A. M. 1922. *The population problem*. Oxford University Press. 516 pp.
10. Chitty, D. 1937. A ringing technique for small mammals. *J. Anim. Ecol.* 6(1): 36–53.
11. Chitty, D. 1952. Mortality among voles (*Microtus agrestis*) at Lake Vyrnwy, Montgomeryshire, in 1936–39. *Philos. Trans.* (B) 236: 505–52, pls. 36–37.
12. Chitty, D. and C. Elton. 1940. The snowshoe rabbit enquiry, 1938–39. *Canad. Field Nat.* 54: 117–24.
13. Chitty, D., and M. Nicholson. 1942. Canadian arctic wildlife enquiry, 1940–41. *J. Anim. Ecol.* 11(2):270–87.
14. Chitty, D., and D. A. Kempson. 1949. Prebaiting small mammals and a new design of live trap. *Ecology* 30 (4): 536–42.
15. Chitty, D., and E. Phipps. 1960. The effect of fleas on spleen size in voles. *Jour. Physiol.* 151: 27–28.
16. Chitty, D., and H. N. Southern, eds. 1958. *Control of rats and mice*. Oxford: Clarendon Press. 3 vols.
17. Chitty, H. 1950. Canadian arctic wildlife enquiry, 1943–49: with a summary of results since 1933. *J. Anim. Ecol.* 19 (2): 180–93.
18. Christian, J. J. 1950. The adreno-pituitary system and population cycles in mammals. *J. Mammal* 31: 247–59.
19. Clarke, J. R. 1953. The effect of fighting on the adrenals, thymus, and spleen of the vole (*Microtus agrestis*). *J. Endocrin.* 9: 114–26.
20. Clarke, J. R. 1955. Influence of numbers on reproduction and survival in two experimental vole populations. *Proc. roy. Soc.* B, 144: 68–85.

21. Collett, R. 1911–12. *Norges Pattedyr. Christiania* 32a-b.
22. Crowcroft, P. 1957. *The life of the shrew.* London: Max Reinhardt. 166 pp.
23. Crowcroft, P. 1966. *Mice all over.* London: Foulis. 123 pp.
24. Crowcroft, P. 1972. The sheep and the saltbush: the utilization of Australia's arid lands. In *The careless technology,* ed. M. Taghi Farvar and J. Milton, 742–52. New York: Natural History Press.
25. Crowcroft, P. 1978. *The zoo.* Sydney: Matthews-Hutchinson. 160 pp.
26. Crowcroft, P., and J. N. R. Jeffers. 1961. Variability in the behaviour of wild house mice (*Mus musculus* L.) towards traps. *Proc. zool. Soc. Lond.* 137: 573–82.
27. Davies, M. 1959. A contribution to the ecology of species of *Notiophilus* and allied Genera (Col.:Carabidae). *Ent. Month. Mag.* 95: 25–28.
28. Davis, D. H. S. 1933. Rhythmic activity in the short-tailed vole, *Microtus. J. Anim. Ecol.* 2 (2): 232–38.
29. Dawson, J. 1956. Splenic hypertrophy in voles. *Nature* Lond. No. 4543: 1183–84.
30. Doty, R. E. 1938. The prebaited feeding-station method of rat control. *Hawaii Plant Rec.* 42: 39–76.
31. Duffey, E. 1953. On a lycosid spider new to Britain and two rare spiders taken near Oxford. *Ann. Mag. Nat. Hist.* (12) 6: 149–57.
32. Duffey, E. 1956. Aerial dispersal in a known spider population. *J. Anim. Ecol.* 25 (1): 85–111.
33. Efford, I. E. 1959. Rediscovery of *Bathynella chappuisi* Delachaux in Britain. *Nature* Lond. 184: 558–59.
34. Efford, I. E. 1960. Observations on the biology of *Tanytarsus (Stempellina) flavidulus* (Edwards) (Diptera, Chironomidae). *Ent. Month. Mag.* 96: 201–3.
35. Efford, I. E. 1962. The taxonomy, distribution and habitat of the watermite, *Feltria romijni* Besseling 1930. *Hydrobiologica* 19: 161–78.
36. Efford, I. E. 1965. Ecology of the watermite *Feltria romijni* Besseling. *J. Anim. Ecol.* 34 (2): 233–51.
37. Elbourn, C. A. 1965. The fauna of a calcareous woodland stream in Berkshire. *Ent. Month. Mag.* 101: 25–30.
38. Elbourn, C. A. 1970. Influence of substrate and structure on the colonization of an artifact simulating decaying oak wood on oak trunks. *Oikos* 21: 32–41.
39. Elton, C. S. 1922. On the colours of water-mites. *Proc. zool. Soc. Lond.* 1922: 1231–39.
40. Elton, C. S. 1924. Periodic fluctuations in the numbers of animals: their causes and effects. *Brit. J. Exper. Biol.* 2: 119–63.
41. Elton, C. 1925. Coleoptera and Lepidoptera from Spitsbergen. Results of the Oxford Expedition to Spitsbergen, 1924. *Ann. Mag. Nat. Hist.* (9) 16: 357–59.
42. Elton, C. 1927. *Animal ecology.* London: Sidgwick and Jackson. 207 pp.
43. Elton, C. 1930. *Animal ecology and evolution.* Oxford University Press. 96 pp.
44. Elton, C. 1931. Epidemics among sledge dogs in the Canadian Arctic and their relation to disease in the Arctic fox. *Can. J. Res.* 5: 673–92.
45. Elton, C. 1933. *Exploring the animal world.* London: Unwin. 119 pp.
46. Elton, C. 1933. The Canadian Snowshoe Rabbit Enquiry, 1931–32. *Can. Fld. Nat.* 47: 63–69, 84–86.

47. Elton, C. 1936. *Bureau of Animal Population, Annual Report.* Oxford: Alden Press. 48 pp.
48. Elton, C. 1942. *Voles, mice and lemmings.* Oxford: Clarendon Press. 496 pp.
49. Elton, C. 1954. An ecological text book. (Review of *Fundamentals of ecology,* by E. P. Odum). *J. Anim. Ecol.* 23 (2): 382–84.
50. Elton, C. 1958. *The ecology of invasions by animals and plants.* London: Methuen. 181 pp.
51. Elton, C. 1966. *The pattern of animal communities.* London: Methuen; New York: John Wiley and Sons. 432 pp.
52. Elton, C. 1966. The value of the small university research institute. *University of Oxford, Commission of Enquiry. IV. Individuals.* Oxford University Press.
53. Elton, C. 1973. The structure of invertebrate populations inside neotropical rain forest. *J. Anim. Ecol.* 42 (1): 55–104.
54. Elton, C., D. H. S. Davis, and G. M. Findlay. 1935. An epidemic among voles (*Microtus agrestis*) on the Scottish border in the spring of 1934. *J. Anim. Ecol.* 4 (2): 277–88.
55. Elton, C., E. B. Ford, J. R. Baker, and A. D. Gardner. 1931. The health and parasites of a wild mouse population. *Proc. zool. Soc. Lond.* 1931: 657–721.
56. Elton, C., and R. S. Miller. 1954. The ecological survey of animal communities: with a practical system of classifying habitats by structural characters. *J. Ecology* 42 (2): 460–96.
57. Elton, C., and M. Nicholson. 1942. The ten-year cycle of the lynx in Canada. *J. Anim. Ecol.* 11 (2): 215–44.
58. Elton, C., et al. 1933. *The Matemak conference on biological cycles.* Abstracts and discussion. Canadian Labrador: Matemak Factory. 50 pp.
59. Emlen, J. T., Jr. 1947. Baltimore's community rat control program. *Amer. J. Publ. Hlth.* 37: 721–27.
60. Emlen, J. T., Jr., A. W. Stokes, and D. E. Davis. 1949. Methods for estimating populations of brown rats in urban habitats. *Ecology* 30: 430–42.
61. Evans, F. C., and W. W. Murdoch. 1968. Taxonomic composition, trophic structure and seasonal occurrence in a grassland insect community. *J. Anim. Ecol.* 37 (1): 259–73.
62. Evans, D. M. 1973. Seasonal variation in the body composition and nutrition of the vole *Microtus agrestis. J. Anim. Ecol.* 42 (1): 1–18.
63. Fager, E. W. 1968. The community of invertebates in decaying oak wood. *J. Anim. Ecol.* 37 (1): 121–42.
64. Gilbert, O., T. B. Reynoldon, and J. Hobart. 1952. Gause's hypothesis: an examination. *J. Anim. Ecol.* 21 (2): 310–12.
65. Godfrey, G. K. 1953. A technique for finding *Microtus* nests. *J. Mammal.* 34 (4): 503–5.
66. Godfrey, G. K. 1954. Tracing field voles (*Microtus agrestis*) with a Geiger-Müller counter. *Ecology* 35 (1): 5–10.
67. Godfrey, G. K., and P. Crowcroft. 1960. *The life of the mole.* London: Museum Press. 152 pp.
68. Graham, E. H. 1944. *Natural principles of land use.* New York: Oxford University Press. 274 pp.
69. Hardy, A. C. 1942. Natural history—old and new. Inaugural address Aberdeen University. *Fishing News,* Aberdeen. 3 pp.

70. Hardy, A. C. 1968. Charles Elton's influence in ecology. *J. Anim. Ecol.* 37(1): 3–8.

71. Hayes, W. J., and T. B. Gaines. 1950. Control of Norway rats with residual rodenticide. *Publ. Hlth. Rep.* Washington, D.C. 65: 1537–55.

72. Hewitt, C. G. 1921. *The conservation of the wild life of Canada.* New York: Scribner's. 344 pp.

73. Howard, W. E. 1949. Dispersal, amount of inbreeding, and longevity in a local population of Prairie deermice on the George Reserve, Southern Michigan. *Contrib. Lab. Vert. Biol. U. Mich.* 43: 1–50.

74. Jenkins, D. 1961. Population control in protected partridges (*Perdix perdix*) *J. Anim. Ecol.* 61 (2): 235–58.

75. Jenkins, D. 1961. Social behaviour in the partridge *Perdix perdix.* *Ibis* 103A: 155–58.

76. Kempson, D. A. 1950. Low-power phase-contrast microscopy without a condenser. *Q. Jour. Micr. Sci.* 91 (1): 109–10.

77. Kempson, D. A. 1952. A new type of light source for phase-microscopy. *Q. Jour. Micr. Sci.* 93 (3): 369–70.

78. Kempson, D. A., and A. Macfadyen. 1954. An inexpensive multipoint recorder for field use. *J. Anim. Ecol.* 23 (2): 376–80.

79. Kempson, D. A., M. Lloyd, and R. Ghelardi. 1963. A new extractor for woodland litter. *Pedobiologia* 3: 1–21.

80. Kikkawa, J. 1959. Habitats of the field mouse on Fair Isle in spring 1956. *Glasgow Nat.* 18: 65–77.

81. Kikkawa, J. 1964. Movement, activity and distribution of the small rodents *Clethrionomys glareolus* and *Apodemus sylvaticus* in woodland. *J. Anim. Ecol.* 33 (2): 259–99.

82. Kitching, R. L. 1971. A core sampler for semi-fluid substrates. *Hydrobiologica* 37: 205–9.

83. Kitching, R. L. 1971. An ecological study of water-filled tree-holes and their position in the woodland ecosystem. *J. Anim. Ecol.* 40 (2): 281–302.

84. Lack, D., and D. F. Owen. 1955. The food of the swift. *J. Anim. Ecol.* 24 (1): 120–36.

85. Larkin, P. A., and C. A. Elbourn. 1964. Some observations on the fauna of dead wood on live oak tress. *Oikos* 15: 79–92.

86. Lewontin, R. 1967, ed. *Population biology and evolution.* Syracuse University Press. 205 pp.

87. Leslie, P. H. 1958. A stochastic model for studying the properties of certain biological systems by numerical methods. *Biometrika* 45 (1–2): 16–31.

88. Leslie, P. H. 1959. The properties of a certain lag type of population growth and the influence of an external random factor on a number of such populations. *Physiol. Zool.* 32 (3): 151–59.

89. Leslie, P. H. 1960. A note on some approximations to the variance in discrete-time stochastic models for biological systems. *Biometrika* 47 (1–2): 196–97.

90. Leslie, P. H. 1962. A stochastic model for two competing species of *Tribolium* and its application to some experimental data. *Biometrika* 49 (1–2): 1–25.

91. Leslie, P. H., and D. Chitty. 1951. The estimation of population parameters from data obtained by means of the capture-recapture method. I. The maximum likelihood equations for estimating the death rate. *Biometrika* 38: 269–92.

92. Leslie, P. H., D. Chitty, and H. Chitty. 1953. The estimation of population parameters from data obtained by means of the capture-recapture method. III. An example of the practical application of the method. *Biometrika* 40: 137–69.
93. Leslie, P. H., and T. Park. 1949. The intrinsic rate of increase of *Tribolium castaneum* Herbst. *Ecology* 30 (4): 469–77.
94. Leslie, P. H., T. Park, and D. B. Mertz. 1968. The effect of varying the initial numbers on the outcome of competition between two *Tribolium* species. *J. Anim. Ecol.* 37 (1): 9–23.
95. Leslie, P. H., and R. M. Ranson. 1940. The mortality, fertility and rate of natural increase of the vole (*Microtus agrestis*) as observed in the laboratory. *J. Anim. Ecol.* 9 (1): 27–52.
96. Leslie, P. H., J. S. Tener, M. Vizoso, and H. Chitty. 1955. The longevity and fertility of the Orkney vole, *Microtus orcadensis*, as observed in the laboratory. *Proc. zool. Soc. Lond.* 125 (1): 115–25.
97. Leslie, P. H., U. M. Venables, and L. S. V. Venables. 1952. The fertility and population structure of the brown rat (*Rattus norvegicus*) in corn-ricks and some related habitats. *Proc. zool. Soc. Lond.* 122: 187–238.
98. Lloyd, M. 1963. Numerical observations on movements of animals between beech litter and fallen branches. *J. Anim. Ecol.* 32 (1): 157–63.
99. Lloyd, M. 1967. "Mean crowding." *J. Anim. Ecol.* 36 (1): 1–30.
100. Lotka, A. J. 1925. *Elements of Physical Biology.* Baltimore: Williams and Wilkins. 460 pp.
101. Macfadyen, A. 1952. The small arthropods of a *Molinia* fen at Cothill. *J. Anim. Ecol.* 21 (1): 87–117.
102. Macfadyen, A. 1961. Improved funnel-type extractors for soil arthropods. *J. Anim. Ecol.* 30 (1): 171–84.
103. Maslow, A. 1971. *The farther reaches of human nature.* Harmondsworth: Penguin. 440 pp.
104. Middleton, A. D. 1936. Factors controlling the population of the partirdge (*Perdix perdix*) in Great Britain. *Proc. zool. Soc. Lond.* 1935 (4)(1936): 795–815.
105. Middleton, A. D., and B. T. Parsons. 1937. The distribution of the Grey squirrel (*Sciurus carolinensis*) in Great Britain in 1937. *J. Anim. Ecol.* 6 (2): 286–90.
106. Miller, R. S. 1954. Food habits of the wood-mouse, *Apodemus sylvaticus* (Schreber, 1780), and the bank vole, *Clethrionomys glareolus* (Schreber, 1780), in Wytham Woods, Berkshire. *Saügetierl. Mitt.* 2 (3): 109–14.
107. Miller, R. S. 1955. Activity rhythms in the wood mouse, *Apodemus sylvaticus* and the bank vole, *Clethrionomys glareolus. Proc. zool. Soc. Lond.* 125 (3–4): 505–19.
108. Miller, R. S. 1958. A study of a wood mouse population in Wytham Woods, Berkshire. *J. Mammal.* 39(4): 477–93.
109. Murdoch, W. W. 1966. Aspects of the population dynamics of some marsh Carabidae. *J. Anim. Ecol.* 35 (1): 127–56.
110. Murdoch, W. W. 1967. Life history patterns of some British Carabidae (Coleoptera) and their ecological significance. *Oikos* 18: 25–32.
111. Newson, J. 1962. Seasonal differences in reticulocyte count, haemoglobin level and spleen weight in wild voles. *Brit. Jour. Haematol.* 8 (3): 296–302.
112. Newson, J., and D. H. Chitty. 1962. Haemoglobin levels, growth and survival in two *Microtus* populations. *Ecology* 44: (4) 733–38.

113. Newson, R. 1963. Differences in numbers, reproduction and survival between two neighbouring populations of bank voles (*Clethrionomys glareolus*). *Ecology* 44 (1): 110–20.
114. O'Connor, J. R. 1948. The use of blood anti-coagulants for rodent control. *Research* London 1 (7): 334–35.
115. Odum, E. P. 1953. *Fundamentals of ecology.* Philadelphia: Saunders. 384 pp.
116. Odum, E. P., and A. J. Pontin. 1961. Population density of the underground ant, *Lasius flavius*, as determined by tagging with P^{32}. *Ecology* 42 (1): 186–88.
117. Paviour-Smith, K. 1960. The fruiting-bodies of macrofungi as habitats for beetles of the Family Ciidae (Coleoptera). *Oikos* 11 (1): 43–71.
118. Paviour-Smith, K. 1960. The invasion of Britain by *Cis bilamellatus* Fowler (Coleoptera:Ciidea). *Proc. roy. Ent. Soc. Lond.* (a) 35: 145–55.
119. Pavoiur-Smith, K. 1968. A population study of *Cis bilamellatus* Wood (Coleoptera:Ciidae). *J. Anim. Ecol.* 37 (1): 205–28.
120. Ranson, R. M. 1934. The field vole (*Microtus*) as a laboratory animal. *J. Anim. Ecol.* 3 (1): 70–76.
121. Ranson, R. M. 1941. Pre-natal and infant mortality in a laboratory population of voles (*Microtus agrestis*). *Proc. zool. Soc. Lond.* A. 111: 45–57.
122. Ratcliffe, F. N. 1938. *Flying fox and drifting sand.* London: Angus and Robertson. 341 pp.
123. Sheail, J. 1976. *Nature in trust.* Blackie: Glasgow and London 270 pp.
124. Shelford, V. E. 1913. *Animal communities in temperate America.* University of Chicago Press.
125. Shorten, M. 1946. A survey of the distribution of the American grey squirrel (*Sciurus carolinensis*) and the British red squirrel (*Sciurus vulgaris leucourus*) in England and Wales in 1944–45. *J. Anim. Ecol.* 15 (1): 82–92.
126. Shorten, M. 1951. Some aspects of the biology of the grey squirrel (*Sciurus carolinensis*). *Proc. zool. Soc. Lond.* 121: 427–59.
127. Shorten, M. 1953. Notes on the distribution of the grey squirrel (*Sciurus carolinensis*) and the red squirrel (*Sciurus vulgaris leucourus*) in England and Wales from 1945 to 1952. *J. Anim. Ecol.* 22 (1): 134–140.
128. Smyth, M. 1966. Winter breeding in woodland mice, *Apodemus sylvaticus* and voles, *Clethrionomys glareolus* and *Microtus agrestis*, near Oxford. *J. Anim. Ecol.* 35 (3): 471–85.
129. Smyth, M. 1968. The effects of the removal of individuals from a population of bank voles *Clethrionomys glareolus*. *J. Anim. Ecol.* 37(1): 167–83.
130. Southern, H. N. 1940. The ecology and population dynamics of the wild rabbit (*Oryctolagus cuniculus*). *Ann. appl. Biol.* 27 (4): 509–26.
131. Southern, H. N. 1970. The natural control of a population of Tawny owls (*Strix aluco*). *J. Zool.* 162: 197–285.
132. Southern, H. N., R. Carrick, and W. G. Potter. 1965. The natural history of a population of Guillemots (*Uria aalge* Port.). *J. Anim. Ecol.* 34(3): 649–65.
133. Southern, H. N., and E. M. O. Laurie. 1946. The House-mouse (*Mus musculus*) in corn ricks. *J. Anim. Ecol.* 37 (1): 75–97.
134. Southern, H. N., and V. P. W. Lowe. 1968. The pattern of distribution of prey and predation in tawny owl territories. *J. Anim. Ecol.* 37 (1): 75–97.

135. Southwick, C. H. 1956. The abundance and distribution of harvest mice (*Micromys minutus*) in corn ricks near Oxford. *Proc. zool. Soc. Lond.* 126 (3): 449–52.

136. Southwick, C. H. 1958. Population characteristics of house-mice living in English corn ricks: density relationships. *Proc. zool. Soc. Lond.* 131 (1): 163–75.

137. Summerhayes, V. S. 1923. Contributions to the ecology of Spitsbergen and Bear Island. *Jour. Ecol.* 11 (2): 214–86.

138. Summerhayes, V. S. 1928. Further contributions to the ecology of Spitsbergen. *Jour. Ecol.* 16 (2): 193–268.

139. Sutton, S. L. 1968. The population dynamics of *Trichoniscus pusillus* and *Philoscia muscorum* (Crustacea, Oniscoidea) in limestone grassland. *J. Anim. Ecol.* 37 (2): 425–44.

140. Todd, V. 1949. The habits and ecology of the British harvestmen (Arachnida, Opiliones), with special reference to those of the Oxford district. *J. Anim. Ecol.* 18 (2): 209–29.

141. Todorovic, M. 1957. The time and moulting pattern of the mole (*Talpa europea* L.). *Ark. Biol. Nauk. Belgrade* 1955 (7) (1957): 47–57.

142. University of Oxford. 1966. *Commission of Enquiry* (Evidence up to 31 Oct. 1964). Oxford University Press.

143. Warwick, T. 1934. The distribution of the muskrat (*Fiber zibethica*) in the British Isles. *J. Anim. Ecol.* 3 (2): 250–67.

144. Warwick, T. 1935. Some escapes of coypus (*Myopotamus coypu*) from nutria farms in Great Britain. *J. Anim. Ecol.* 4 (1): 146–47.

145. Warwick, T. 1941. A contribution to the ecology of the musk-rat (*Ondatra zibethica*) in the British Isles. *Proc. zool. Soc. Lond.* A. 110: 165–201.

146. Watson, J. S. 1951. *The rat problem in Cyprus.* Colonial Res. Publ. 9. London: HMSO. 66 pp.

147. Watts, C. H. S. 1968. The foods eaten by wood mice (*Apodemus sylvaticus*) and bank voles (*Clethrionomys glareolus*) in Wytham Woods, Berkshire. *J. Anim. Ecol.* 37 (1): 25–41.

148. Watts, C. H. S. 1970. Effects of supplementary food on breeding in woodland rodents. *J. Mammal.* 51: 169–71.

149. Wells, A. Q. 1937. Tuberculosis in wild voles. *The Lancet,* May 2, 1937: 1221.

150. Wells, A. Q. 1938. The susceptibility of voles to human and bovine strains of tubercle bacilli. *Brit. Jour. exp. Path.* 19: 324–28.

151. Whittaker, J. B. 1969. Quantitative and habitat studies of the frog-hoppers and leaf-hoppers (Homoptera, Auchenorhyncha) of Wytham Woods, Berkshire. *Ent. Month. Mag.* 105: 27–37.

152. Williams, G. R. 1954. Population fluctuations in some northern hemisphere game birds (Tetraonidae). *J. Anim. Ecol.* 23 (1): 1–37.

Index